McGRAW-HILL

SCIENCE

Macmillan/McGraw-Hill Edition

Reading in Science Workbook

GRADE 3

Macmillan McGraw-Hill
New York Farmington

1 (t)Alena Brozova/Hemera/Getty Images, (c)©Image Source, (b)Glow Images; **32** ranplett/Getty Images; **49** (t)Comstock Images/Alamy, (b)Comstock Images/Alamy; **99** (r)Robert Madden/National Geographic/Getty Images, (l)UUSGS Photograph by Harry Glicken; **113** (cl)Joyce Photographics/Science Source, (cr)Siim Sepp/Alamy, (br)Juniors/Juniors/Superstock, (bl)©Siim Sepp/Alamy, (tr)Jacques Cornell photographer/McGraw Hill Education, (tl)©Photolibrary/age fotostock; **114** (l)McGraw-Hill Education, (r)Digital image courtesy of the Getty's Open Content Program, (c)Ryan McVay/Photodisc/Getty Images; **120** (l)Ken Cavanagh Photographer/McGraw Hill Education, (r)Jacques Cornell photographer/ McGraw-Hill Education, (c)Ken Cavanagh Photographer/McGraw-Hill Education; **125** Marvin Dembinsky Photo Associates/Alamy; **138** (tc)Getty Images, (t)broken3/Getty Images, (bc)Martin Diebel/Getty Images, (b)©Mark Dierker, photographer/McGraw-Hill Education; **156** Camden Yehle/WeatherVideoHD.TV; **161** (r)Art Montes De Oca/The Image Bank/Getty Images, (l)Stefan Witas/Getty Images; **181** (l)Jeff Greenberg/Photolibrary/Getty Images, (r)©Lou Paintin/Getty Images; **187** (br)Charles D Winters/Science Source/Getty Images, (bl)Mike Clarke/Getty Images, (tl)©Jan Tadeusz/Alamy, (tr)Kristian Husar/Getty Images; **205** (c)Somchaisom/Getty Images, (r)Lio22/iStock/Getty Images, (l) Gary Dyson/Alamy; **206** (l)NASA, (r)Image Science and Analysis; **211** (l)USGS Astrogeology Science Center, (cr)Ingram Publishing, (r)Steve Lee (University of Colorado), Jim Bell (Cornell University), Mike Wolff (Space Science Institute), and NASA, (cl)Science Source; **212** (l)©Bettmann/Corbis, (b)NASA Jet Propulsion Laboratory (NASA-JPL), (r)NASA-JPL, (cl)NASA/JPL, (cr)NASA/JPL/STScI; **225** (t)Alan and Sandy Carey/Getty Images, (r)Mike Hill/Getty Images, (b)Alan and Sandy Carey/Getty Images, (l)Heinrich van den Berg/Getty Images; **231** (tl)McGraw-Hill Education, (cl)McGraw-Hill Education, (cr)McGraw-Hill Education, (r)McGraw-Hill Education, (tr)McGraw-Hill Education; **300** (l)Terje Rakke/The Image Bank/Getty Images, (r)Ryerson Clark/Getty Images, (c)Glow Images; **Back Cover** (t)Studio-Annika/iStock/GettyImages, (r)Studio-Annika/iStock/GettyImages; **Front Cover** (r)ARCO/P. Wegner/age fotostock, (l)sserg_dibrova/iStock/Getty Images; **title page** (l)sserg_dibrova/iStock/Getty Images, (r)Studio-Annika/iStock/GettyImages.

Macmillan/McGraw-Hill

A Division of The McGraw-Hill Companies

Published by Macmillan/McGraw-Hill, of McGraw-Hill Education, a division of The McGraw-Hill Companies, Inc., Two Penn Plaza, New York, New York 10121.

Printed in the United States of America
22 23 24 25 26 27 QVS 19 18 17 16 15

Roles for Plants and Animals

Fill in the blanks. Reading Skill: **Summarize** - questions 1–4, 10–12

How Do Living Things Use Air?

1. Plants give off ________________, a gas that animals need.
2. Animals take in the oxygen made by ________________.
3. Animals ________________ carbon dioxide, a gas that plants need.
4. Plants take in carbon dioxide given off by ____________.
5. The carbon dioxide and oxygen cycles are the process of trading ________________________.
6. Animals breathe in ________________ and breathe out ________________.
7. Plants take in ________________ and give off ________________.
8. The world's largest oxygen-producing organisms are ____________.
9. The most important source of oxygen is ________________, which make more oxygen than all the land plants in the world.

Name ______________________ Date ____________

Lesson Outline

Lesson 3

Fill in the blanks.

How Do Populations Depend on Each Other?

10. Animals that hunt for food are called ______________.

11. The animals that predators eat are called ______________.

12. Animals that eat dead animals are called ______________.

How Can Populations Affect Each Other?

13. The clownfish is coated with slime that ______________ it from the stingers of the ______________.

14. The anemone depends on the clownfish for ______________.

15. Cattle egrets follow cattle to get ______________.

16. Cattle are neither helped nor harmed by ______________.

17. An organism that lives on or inside another organism is called a(n) ______________.

18. The organism a parasite lives with is called the ______________.

19. Fleas are parasites on ______________.

How Do Animals Help Plants Reproduce?

20. When acorns fall close to oak trees, there is not enough ______________ for the seeds to grow.

21. Squirrels help oak trees reproduce by ______________ acorns far from the trees.

Name________________________________ Date______________

Interpret Illustrations
Lesson 3

How Do Living Things Use Air?

A diagram uses pictures and words to describe a thing or a process. This diagram shows the carbon dioxide and oxygen cycles. Notice how the cycles are based on the needs of plants and animals. If you follow the arrows, you can see how the gases are passed from one group to another.

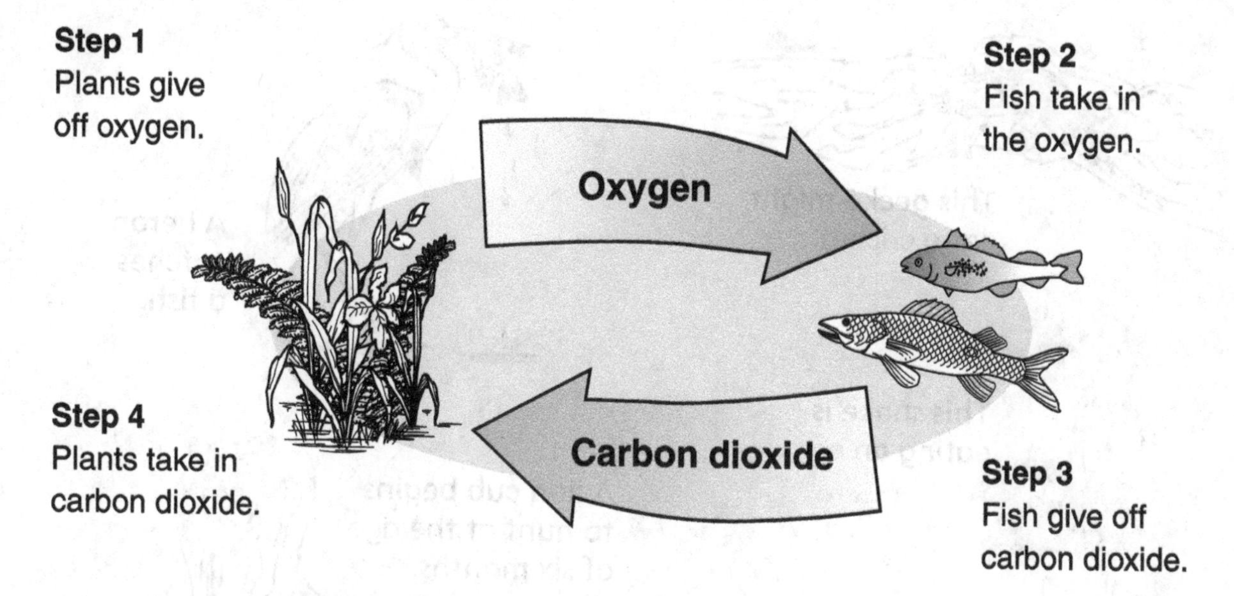

Use the above diagram to fill in the blanks.

1. In step 1, plants give off ________________.
2. The oxygen made by plants is taken in by ________________.
3. Fish release ________________.
4. The carbon dioxide released by the fish is taken in by ________________.
5. Plants make the ________________ that fish need, while fish make the ________________ that plants need.

Name ______________________ Date ____________

Interpret Illustrations

Lesson 3

How Do Populations Depend on Each Other?

Animals that are hunters are called predators. Animals that are hunted and eaten by predators are prey.

This gecko might eat a cricket.

A heron catches a fish.

This snake is eating an egg.

A lion cub begins to hunt at the age of six months.

Answer these questions about the diagram above.

1. Which animals are the predators in the illustrations?

____

____

2. Which animals are the prey in the illustrations?

____

____

3. How can you tell which animals are the predators?

____

____

Name______________________________ Date__________

Lesson Vocabulary
Lesson 3

Roles for Plants and Animals

Vocabulary

- carbon dioxide
- host
- parasite
- oxygen
- carbon dioxide and oxygen cycles
- scavenger
- prey
- predators

Fill in the blanks.

1. The process of trading oxygen and carbon dioxide is called the ______________________________.
2. Plants give off the gas ________________.
3. Animals give off ________________.
4. An organism that lives on or inside another organism is called a(n) ________________.
5. An organism that has another organism living inside or on it is called a(n) ________________.
6. Animals that hunt other animals for food are ________________.
7. Predators hunt and catch ________________.
8. A ________________ eats dead animals that it did not kill.

Look at the pairs of organisms below. On each line write the letter of the term that describes their relationship.

a. both are helped
b. one is helped and one is not
c. one is helped and the other is harmed

9. squirrels and oak trees _____
10. sea anemones and clownfish _____
11. cattle egrets and cows _____
12. fleas and dogs _____
13. tapeworms and humans _____

Name ______________________ Date ____________

Cloze Test

Lesson 3

Roles for Plants and Animals

Vocabulary

parasite
animal
hosts
food
blood

Fill in the blanks.

An organism that lives on or inside a host is a(n) ______________.
The host of a tapeworm can be a(n) ______________ or a human. After the tapeworm attaches itself to the host, it takes in ______________ that the host has digested. Parasites usually harm their ______________, sometimes killing them. Fleas eat the ______________ of dogs as food.

Name____________________ Date__________

Chapter Vocabulary

Chapter 3

Relationships Among Living Things

Circle the letter of the best answer.

1. An organism that eats other organisms is called a
   a. consumer. b. decomposer.
   c. parasite. d. producer.

2. Which of the following is a diagram that shows how energy moves through an ecosystem?
   a. carbon dioxide and oxygen cycle b. energy pyramid
   c. food chain d. food web

3. All the members of a single type of living thing in an area describes a
   a. community. b. food web.
   c. niche. d. population.

4. An organism that a parasite lives on or inside is called a
   a. consumer. b. decomposer.
   c. host. d. producer.

5. All the living things in an ecosystem describes a
   a. community. b. food web.
   c. niche. d. population.

6. An organism that breaks down dead plant and animal material is a
   a. consumer. b. decomposer.
   c. parasite. d. producer.

Name ____________________ Date __________

Chapter Vocabulary

Chapter 3

Circle the letter of the best answer.

7. Which of the following best describes the place where a plant or animal naturally lives and grows?
   a. community b. energy pyramid
   c. food web d. habitat

8. An organism that lives on or inside another organism is a
   a. consumer. b. decomposer.
   c. parasite. d. producer.

9. Which of the following describes all the living and nonliving things in an environment?
   a. ecosystem b. energy pyramid
   c. food web d. population

10. A series of organisms that depend on one another for food is a
   a. community. b. energy pyramid.
   c. food chain. d. population.

11. Which of the following describes an organism that makes its own food?
   a. consumer b. decomposer
   c. host d. producer

12. Which of the following describes several food chains that are connected?
   a. community b. energy pyramid
   c. food web d. population

Name____________________ Date__________

Chapter Summary

Chapter Summary

1. What is the name of the chapter you just finished reading?

____

____

2. What are two vocabulary words you learned in the chapter? Write a definition for each.

____

____

____

3. What are two main ideas that you learned in this chapter?

____

____

____

____

____

____

Name________________________________ Date______________

Chapter Graphic Organizer
Chapter 4

Ecosystems in Balance

Comparing and Contrasting from a Table

Below is a list of features that both a rain forest ecosystem and temperate forest ecosystem have. Compare and contrast the two systems by putting a check in the column of the ecosystem that has more of the feature listed. Answer the questions at the bottom of the page.

	Rain Forest	Temperate Forest
kinds of trees	( )	( )
types of birds	( )	( )
rain	( )	( )
types of ants	( )	( )
types of plants	( )	( )
types of animals	( )	( )
heat	( )	( )

What do the two ecosystems have in common?

____

____

____

In what ways are they different? Why is this important?

____

____

____

Name________________________________ Date______________

Chapter Reading Skill
Chapter 4

Compare and Contrast

Remember, to compare things, you find how they are alike. To contrast things, you find how they are different. If a puzzle below is labeled **compare,** circle the two objects that are exactly alike. If it's labeled **contrast,** circle the one object that's different.

Compare

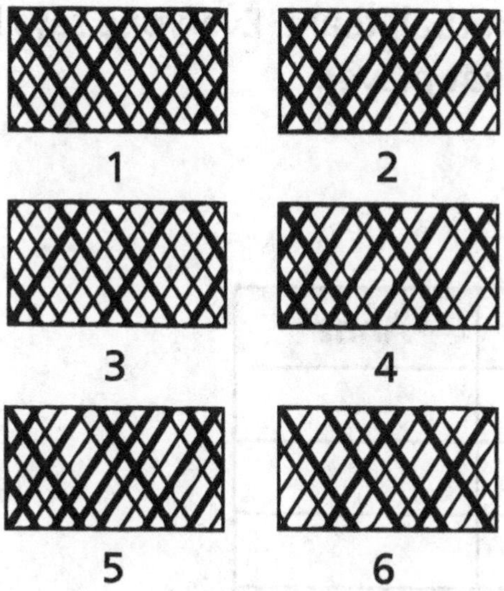

1 2
3 4
5 6

Contrast

1 2
3 4
5 6

Contrast

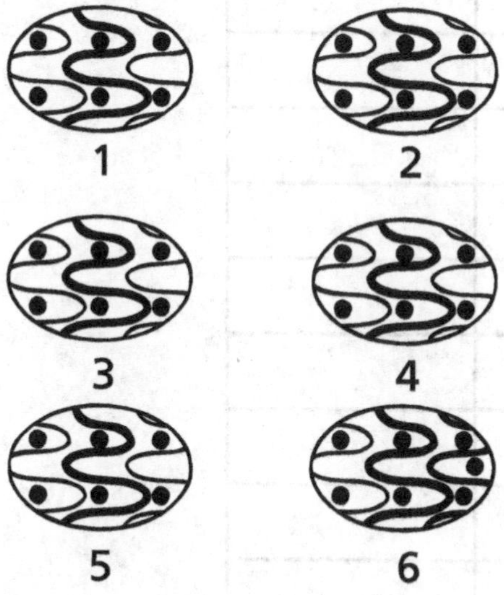

1 2
3 4
5 6

Compare

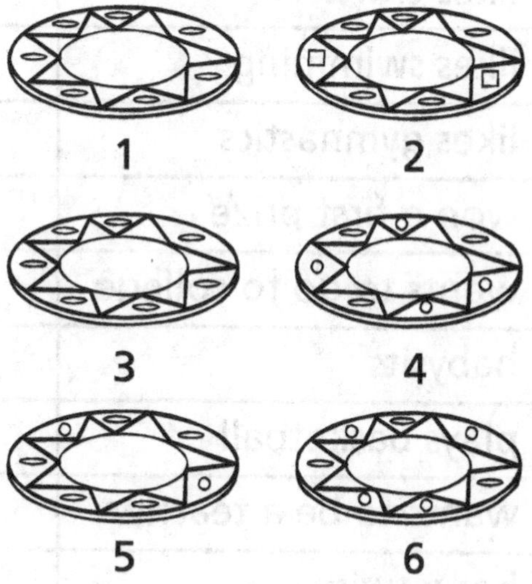

1 2
3 4
5 6

Name ______________________ Date ______________

Chapter Reading Skill
Chapter 4

Alike, Yet Different

Read the following story. Then use the chart to compare and contrast the characters.

Carlo and his sister Anna are 10-year-old twins. They are both on the camp swimming team. Carlo also likes crafts. Last year he won first place for a lamp he made!

Anna spends lots of time in gymnastics. Last year she won first prize in the tumbling tourney!

Anna likes to baby-sit. She wants to go to college and study to be a teacher. Carlo hopes that when he gets to college, he'll be on the basketball team. Everyone says he's a great player.

	Carlo	Anna
1. is 10 years old		
2. is a boy		
3. is a girl		
4. goes to camp		
5. likes crafts		
6. likes swimming		
7. likes gymnastics		
8. won a first prize		
9. wants to go to college		
10. babysits		
11. plays basketball		
12. wants to be a teacher		
13. has a twin		

Name________________________ Date____________

Lesson Outline
Lesson 4

Competition Among Living Things

Fill in the blanks. Reading Skill: **Compare and Contrast** - questions 9, 10

How Much Room Do Organisms Need?

1. When one organism works against another to get what it needs to live, it is called ________________.
2. Organisms may compete for:
   a. ________________,
   b. ________________,
   c. ________________, or some other need.
3. Desert plants compete for ________________.
4. Animals that hunt for food are called ________________.
5. Animals that predators hunt are called ________________.
6. Predators compete for ________________.
7. Different ecosystems support different numbers of ________________.
8. The most common type of forests in the United States are ________________ forests.
9. Rain forests support many more kinds of trees and birds than ________________ forests.
10. There are many more organisms in rain forests because rain forests are ________________ than temperate forests.

Name ______________________ Date ____________

Lesson Outline

Lesson 4

Fill in the blanks.

Can Competition Be Avoided?

11. Competition is a struggle for ____________.

12. The job or role of an organism in an ecosystem is its ____________.

13. An organism's niche includes

a. ______________________,

b. ______________________, and

c. ______________________.

14. There are many types of pigeons in the forests of New Guinea. Each type has a different ____________.

Name______________________________ Date____________

Interpret Illustrations
Lesson 4

Comparing Forests

Look at the chart recording the number of birds and trees in two kinds of forests. Compare the number of birds and trees in the two forests.

	Temperate forest	**Rain forest**
Different types of trees	 50 to 60	 500 to 600
Different types of birds	 50	 250

Fill in the blanks.

1. What are the two types of forests compared in the chart?
   ____________________ and ____________________
2. How many different kinds of trees does the temperate forest have?
   ____
3. How many different kinds of trees does the rain forest have?
   ____
4. How many different kinds of birds does the temperate forest have?
   ____
5. How many different kinds of birds does the rain forest have? ________

Name________________________ Date______________

Interpret Illustrations

Lesson 4

The Niche of a Victoria Crowned Pigeon

The Victoria crowned pigeon is one kind of pigeon found in New Guinea. It shares an ecosystem with other kinds of pigeons. Study the illustration below. Think about the crowned pigeon's niche in its ecosystem.

Use the above illustration to fill in the blanks.

1. What type of food does the crowned pigeon eat?

____

____

2. Where might the crowned pigeon find this kind of food?

____

____

3. Where does the crowned pigeon make its home?

____

____

Name ______________________ Date ____________

Lesson Vocabulary
Lesson 4

Competition Among Living Things

Vocabulary
competition
niche

Fill in the blanks.

1. An organism's job or role in an ecosystem is called its ______________.
2. When one organism works against another to get what it needs to live, ______________ occurs.

Use the following terms to fill in the blanks below: food, water, space.

3. Plants that live in the desert compete for ______________.
4. Rabbits in a field of grass compete for ______________.
5. Plants that grow in a rain forest compete for ______________.
6. Birds who live in a big city compete for ______________.
7. Describe niches for the following people in your school: you, your teacher, the nurse.

____

____

____

8. Why do more plants grow in the rain forest than in the temperate forest?

____

____

____

Name ______________________________ Date ______________

Cloze Test

Lesson 4

Competition Among Living Things

Vocabulary

water	**competition**
food	**space**
cactus	**plants**

Fill in the blanks.

Whenever one organism works against another to get what it needs to live, ________________ occurs. Organisms may compete for water, ________________, or ________________. Plants in the desert compete for ________________. A ________________ soaks up all of the moisture in a single area. No other ________________ can grow in this area.

Name________________________________ Date______________

Lesson Outline
Lesson 5

Adaptations for Survival

Fill in the blanks.  Reading Skill: **Compare and Contrast** - questions 6, 7, 15, 19, 20

What Is an Adaptation?

1. Parts of organisms work like ________________.
2. A bird uses its ________________ as a tool for eating.
3. Different beak shapes are suited to different kinds of ______________.
4. A special characteristic that helps an organism survive is a(n) ________________.
5. The beak shapes of honeycreepers are adaptations that help the birds ________________.
6. One kind of honeycreeper has a long, curved beak, good for eating ________________ from flowers.
7. A short, thick, and strong beak is good for eating ______________, while a straight beak is good for eating ________________.
8. Most organisms have a variety of ________________.
9. The wool of a lamb is an adaptation designed to keep the lamb ________________.
10. Frogs have long, sticky tongues and ________________ to help them catch insects.
11. The bright coloring of flowers help them survive by attracting insects that help the flowers ________________.

Name______________________________ Date______________

Fill in the blanks.

How Can Adaptations Protect Living Things?

12. Not all adaptations are important for getting food or ______________.

13. Many adaptations help protect an organism from ______________.

14. An adaptation that allows an organism to blend in with its environment is ______________.

15. A white rabbit blends in with the ______________, while brown rabbits match their ______________ habitat.

16. When an animal looks like another animal, it is called ______________.

What Do Animals Do to Defend Themselves?

17. Animals defend themselves by using their ______________.

18. Some animals defend themselves by fighting with ______________, ______________, and ______________.

How Do Animals in Different Environments Adapt?

19. Wolves in the tundra have ______________, while wolves that live in the desert have ______________ fur.

20. The snowy owl has ______________, and the polar bear has ______________.

21. There are ______________ in the cold arctic tundra.

22. The road runner has brown feathers to blend into the ______________ environment.

Name______________________________ Date______________

How Does the Shape of a Bird's Beak Affect What It Eats?

Look at the chart. Notice how each type of honeycreeper has a different beak shape from the other birds. Each beak shape is an adaptation. These adaptations help the honeycreepers to survive.

Honeycreepers

Each type of honeycreeper has one of these three basic beak shapes.

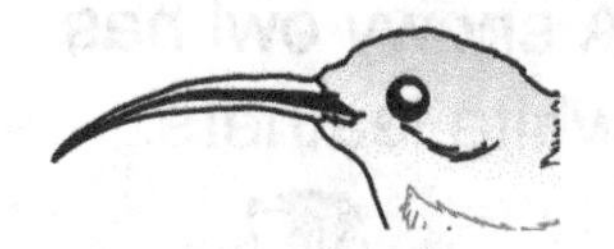

A long, curved beak is good for eating nectar from flowers.

A beak that is short, thick, and strong is just right for eating seeds and nuts.

A straight beak is good for eating insects.

Fill in the blanks.

1. The honeycreepers shown in the chart are alike except for their ______________ shapes.
2. The beaks are adaptations that are related to the way each bird ______________.
3. A long, curved beak is suited to sip ______________ from flowers.
4. Compare the beak of the seed-eating honeycreeper with the beak of the insect-eating honeycreeper.

____

____

____

Name____________________________ Date______________

How Do Animals in Different Environments Adapt?

The conditions in each environment result in adaptations in its living things. The chart lists the adaptations of three different organisms, in three different environments.

Adaptations in Different Environments

	Trees	Bears	Birds
Arctic tundra	Only very small trees grow in the tundra.	A polar bear has thick white fur.	A snowy owl has white feathers.
Desert	A mesquite tree has deep roots.	No bears live in the desert.	A roadrunner has brown feathers.

Use the above chart to fill in the blanks.

1. In which environment is the mesquite tree found? ______________
2. How is the mesquite tree adapted to live in its environment?

   ____
3. In which environment do only small trees live? ______________
4. Do bears live in the desert? ______________
5. Which kinds of bird is adapted to live in the tundra? ______________

Name____________________ Date__________

Lesson Vocabulary

Lesson 5

Adaptations for Survival

Use the word "adaptation," "camouflage," or "mimicry" to fill in the blanks below.

1. Allows organisms to blend in with their environments ______________
2. Helps an organism survive in its environment ______________
3. Organisms that look like other organisms ______________
4. Brown fur on an animal in a forest ______________
5. A long, sticky tongue on a frog ______________
6. The shape of a bird's beak ______________
7. An eyespot on a butterfly ______________
8. A giraffe's long neck ______________
9. A rabbit's white fur ______________
10. A caterpillar looks like a snake ______________
11. A flower's bright color ______________
12. Compare the adaptations of birds that live in the desert and the arctic tundra. Explain how each adaptation helps the organism survive.

____

____

____

____

____

Name ______________________ Date ____________

Cloze Test
Lesson 5

Adaptations for Survival

Vocabulary

beaks	survive
flower	warm
neck	adaptation

Fill in the blanks.

Different types of honeycreepers have ________________ that are shaped differently. Each shape is an ________________, or a special characteristic that helps an organism ________________ in its environment. Wool keeps a lamb ________________ during winter days. A giraffe's long ________________ helps it find food in high places. The bright coloring of a(n) ________________ attracts insects.

Name________________________________ Date______________

Lesson Outline
Lesson 6

Changing Ecosystems

Fill in the blanks. Reading Skill: **Compare and Contrast** - question 20

What Happens When Ecosystems Change?

1. When an ecosystem ____________________, the organisms living there are affected.
2. After the eruption of Mt. St. Helens, the forest was buried in ____________________ that hardened into a tough crust.
3. Organisms that made their homes on or near trees were ____________________.
4. Organisms near the ground had their habitats ____________________.
5. Besides volcanic eruptions, floods and droughts also change a(n) ____________________.

How Do Ecosystems Come Back?

6. In order for a forest to come back after a fire, it must go through several ____________________.
7. When a habitat is destroyed, ____________________ may survive.
8. The grasses that cover the bare ground add ____________________ to the soil.
9. The grasses are replaced by ____________________ that block out light.
10. As the trees grow larger, ____________________ is blocked from reaching the forest floor, causing the grasses to die.

Name ______________________ Date __________

Lesson Outline
Lesson 6

11. Organisms respond to change by:

a. ______________,

b. ______________, or

c. ______________.

12. The fireweed on Mount St. Helens adjusted to its new habitat by growing through the ______________.

13. When organisms do not survive, they ______________.

14. When an organism finds a new home, it ______________.

Are Living Things Dying Out?

15. Threats to organisms include:

a. ______________,

b. ______________,

c. ______________, and

d. ______________.

16. An organism that has very few of its kind left is ______________.

Have Living Things Died Out?

17. Endangered organisms may become ______________.

18. Extinct means that there are ______________ of that type of organism alive.

19. Extinct organisms include the ______________ and ______________.

20. Extinct organisms are gone forever, while ______________ ______________ have very few of their own kind still alive.

Name_________________________ Date______________

Interpret Illustrations

Lesson 6

What Happens When Ecosystems Change?

The illustrations below show Mt. St. Helens before and after it erupted. Study the illustrations. Think about how the ecosystems changed.

Answer these questions about the illustrations above.

1. Which picture shows Mt. St. Helens before it erupted? How do you know?

____

2. How did the animal habitats change after the eruption of Mt. St. Helens?

____

____

3. What happened to the plant life after the volcano erupted?

____

____

4. Some animals survived the eruption. What do you think happened to these animals?

____

____

Name ______________________ Date ____________

Interpret Illustrations

Lesson 6

How Do Ecosystems Come Back?

The diagram shows the changes a forest must go through in order to return to its original condition after a fire. Notice that the pictures are numbered to show the order in which changes take place.

Stage 1 Habitat destruction
Bulbs and seeds may survive underground. They begin to grow in the ash.

Stage 2 Grasses
Over time, grasses cover the bare ground. The grasses add nutrients to the soil. They also provide a home for insects. The insects attract larger animals.

Stage 3 Larger plants
Small trees begin to grow. The trees block the sunlight. Without light the grasses begin to die.

Stage 4 Forest
Small trees are replaced by larger trees. The forest is the final stage.

Use the above diagram to fill in the blanks.

1. In the first stage, bulbs and seeds that have survived the fire begin to grow in the ______________.
2. When grasses return to cover the bare ground, they add ______________ to the soil.
3. The grasses provide a home for ______________.
4. The insects that return to the forest attract ______________.
5. Trees block the Sun causing the ______________ to die.

Name ______________________ Date ____________

Lesson Vocabulary

Lesson 6

Changing Ecosystems

Vocabulary

extinct

perish

endangered

relocate

Fill in the blanks.

1. To fail to survive is to ______________.
2. When there is no more of that type of organism alive it is said to be ______________.
3. To ______________ means to find a new home.
4. When there are very few of its kind left, an organism becomes ______________.

In the space provided, describe what happens during each phase of a forest's return after a fire.

5. Stage 1: Habitat destruction

____

6. Stage 2: Grasses

____

7. Stage 3: Larger plants

____

8. Stage 4: Forest

____

Answer the following question in the space provided.

9. Name three things that can make an ecosystem change.

____

____

____

Name ______________________ Date ______________

Cloze Test
Lesson 6

Changing Ecosystems

Vocabulary

extinct	adjusting	endangered
change	relocate	survive

Fill in the blanks.

Organisms respond to a(n) ______________ in their habitat in one of three ways. Some organisms respond to change in their habitat by ______________. Others perish, or do not ______________. Some organisms ______________, or find a new home. An organism that has very few of its kind left is ______________. An endangered organism is in danger of becoming ______________, meaning that there are no more of that type of organism alive.

Name________________________ Date____________

Chapter Vocabulary
Chapter 4

Ecosystems in Balance

Circle the letter of the best answer.

1. Something that helps an organism survive in its environment is called a(n)

a. adaptation. **b.** characteristic.
c. extinction. **d.** reproduction.

2. One organism working against another to get what it needs to live is called

a. adaptation. **b.** camouflage.
c. competition. **d.** extinction.

3. A hunted animal is called

a. competitor. **b.** endangered.
c. predator. **d.** prey.

4. An animal that may become extinct is

a. adapted. **b.** camouflaged.
c. endangered. **d.** temperate.

5. An animal that hunts other animals for food is called a(n)

a. adapter. **b.** competitor.
c. predator. **d.** prey.

6. An organism that does not survive is said to

a. adapt. **b.** compete.
c. perish. **d.** relocate.

Name ______________________ Date ____________

Chapter Vocabulary

Chapter 4

Circle the letter of the best answer.

7. Which of the following describes when there are no more of that type of organism alive?
   a. adaptation b. competition
   c. extinction d. relocation

8. An organism's job or role in an ecosystem is called its
   a. adaptation. b. camouflage.
   c. competition. d. niche.

9. An adaptation that allows an organism to blend in with its environment is called
   a. camouflage. b. competition.
   c. extinction. d. relocation.

10. To find a new home means to
   a. adapt. b. compete.
   c. perish. d. relocate.

11. A brightly colored flower attracting insects is an example of
   a. adaptation. b. camouflage.
   c. competition. d. a niche.

12. Which of the following is NOT a strategy for survival?
   a. adaptation b. camouflage
   c. competition d. extinction

Name ______________________ Date __________

Unit Vocabulary
Unit B

Choose-a-Word

Circle the word that best completes each sentence.

1. All the living things and nonliving things in a place make up an __________.

   adaptation **atmosphere** **ecosystem**

2. A __________ breaks down dead plants and animals.

   consumer **decomposer** **producer**

3. An organism that lives on or inside another organism is a __________.

   parasite **community** **producer**

4. A __________ hunts for other animals.

   producer **predator** **prey**

5. To __________ means to find a new home.

   perish **adapt** **relocate**

6. There's __________ when more than one organism wants the same thing.

   cooperation **competition** **help**

7. When an animal's __________, there are no more.

   plentiful **adapted** **extinct**

8. A parasite lives on or inside a __________.

   host **niche** **comet**

9. With __________, an animal can hide in plain sight.

   motion **camouflage** **energy**

Name ______________________ Date ____________

Unit Vocabulary

Unit B

Using Words

Vocabulary

mimicry	decomposer	producer	niche	perish
extinct	competition	community	predator	consumer

Read each question below. Write your answer on the line. Use a word from the box above.

1. Which word includes all the living things in an ecosystem?

   ____

2. Which word means "to not survive"? ______________
3. Which word describes one organism working against others to get what it needs? ______________
4. Which word describes an organism that makes its own food?

   ____

5. Which word describes an animal that hunts other animals?

   ____

6. Which word describes an organism's job or role in an ecosystem?

   ____

7. Which word describes an organism that breaks down dead plants and animals? ______________
8. Which word describes an organism that eats producers?

   ____

9. Which word describes an adaptation where animals look like another organism? ______________
10. Which word describes an organism of which there are no more of that type alive? ______________

Word Webs

A word web lists words that describe or relate to the same thing. Here's an example using the word *prey*.

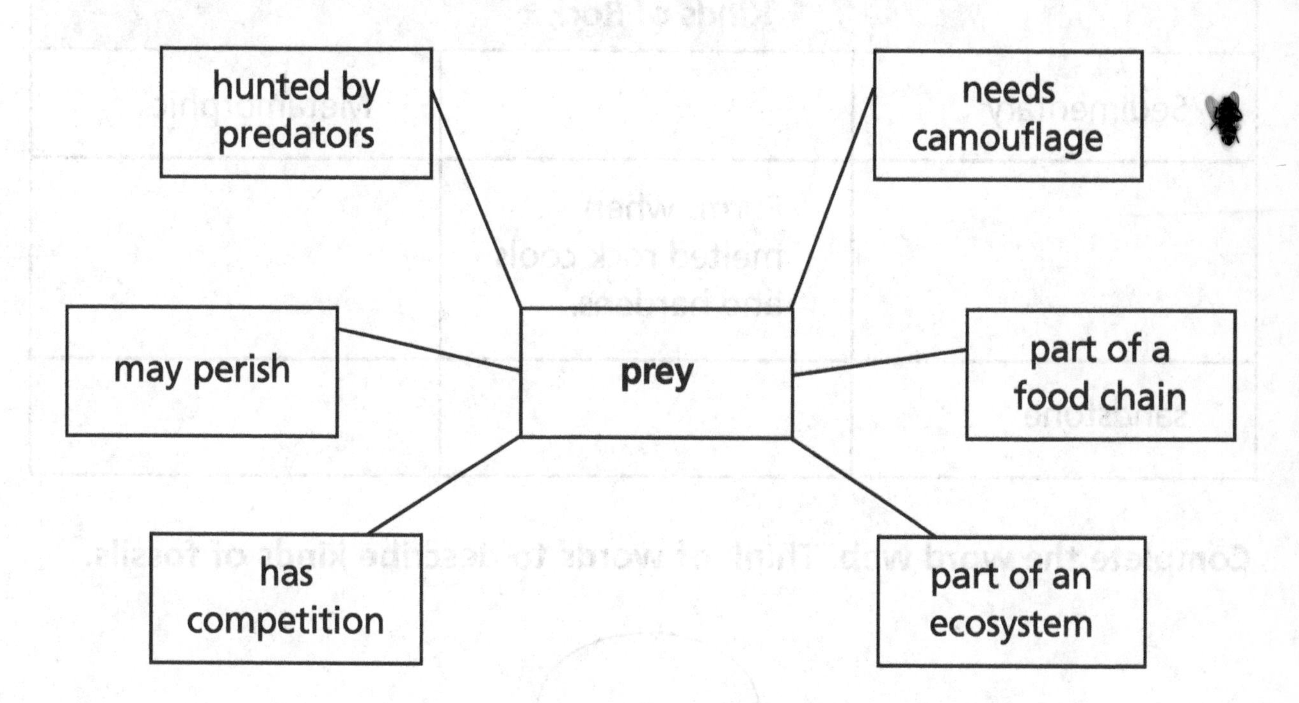

Now it's your turn. Choose a vocabulary word or phrase, such as *community, food web, ecosystem, camouflage, energy pyramid,* or *niche*. Write other words that relate to or describe it.

Name________________________________ Date______________

Chapter Graphic Organizer
Chapter 5

Earth's Resources

Complete the chart.

Kinds of Rock		
Sedimentary		Metamorphic
	Forms when melted rock cools and hardens.	
sandstone		

Complete the word web. Think of words to describe kinds of fossils.

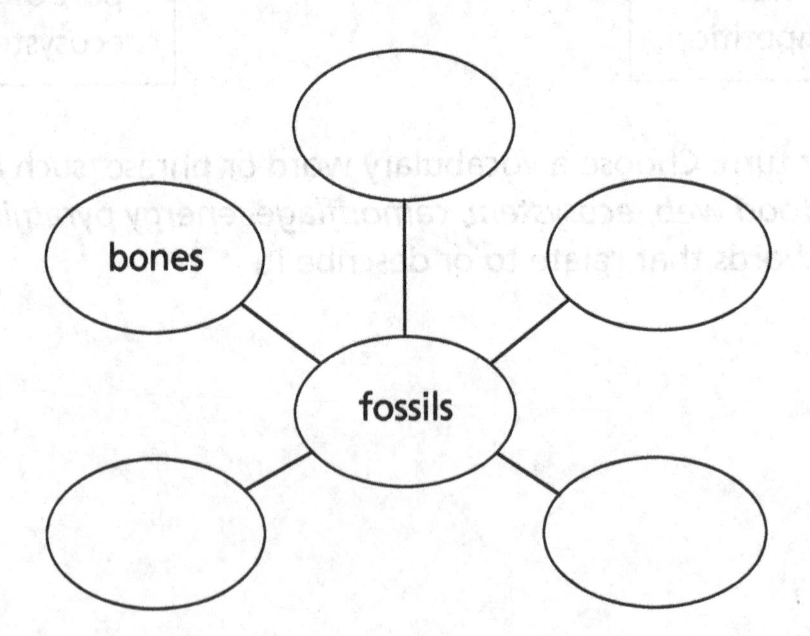

Name_______________________________ Date____________

Chapter Reading Skill

Chapter 5

Sequence of Events

Everything happens in sequence, step by step. **First** you are a baby, **next** you're a child, **then** you'll be a teenager, and **finally** you'll be an adult. You can't be 13 years old **before** you're 10 years old, and you can't be 11 years old **after** you're 13!

Did you notice all the time words in the first paragraph? Those words tell you things are happening in step-by-step sequence.

Read each list below. Write 1, 2, 3, 4, and so on to show the step-by-step sequence.

_____ Rinse hair.	_____ Let the cake cool in the pan.
_____ Get out of tub carefully.	_____ Bake until brown on top.
_____ Fill the tub about half full.	_____ Put cake pan in the oven.
_____ Get soap and shampoo.	_____ Mix until smooth.
_____ Wash body.	_____ Cut a slice of cake, and eat it.
_____ Turn on the warm water.	_____ Pour batter into cake pan.
_____ Dry hair and body.	_____ Spread icing on cake.
_____ Get into tub carefully.	_____ Put cake mix, egg and water in large bowl.
_____ Shampoo hair.	_____ Remove cake from pan.
_____ Get a washcloth and towel.	_____ Get a large bowl and spoon.
_____ Let water out of tub.	

Name________________________________ Date______________

Chapter Reading Skill
Chapter 5

Schedule Sequence

It's important to be able to follow a sequence of events. The schedule for your school day is in sequence, going from one activity to another. Traveling is also something we do in sequence. We even have to plan trips to fit the sequence of buses, trains, planes, and boats!

Look at the bus schedule below. It lists trips between York and Lincoln. Use information from the schedule to answer the questions.

Bus Schedule

Monday through Friday

Leaves	**Arrives**
York 7 A.M.	Lincoln 9 A.M.
Lincoln 10 A.M.	York noon
York 3 P.M.	Lincoln 5 P.M.
Lincoln 6 P.M.	York 8 P.M.

Saturday and Sunday

Leaves	**Arrives**
York 9 A.M.	Lincoln 11 A.M.
Lincoln noon	York 2 P.M.
York 5 P.M.	Lincoln 7 P.M.
Lincoln 8 P.M.	York 10 P.M.

1. At what time Thursday morning does a bus leave York?

____

2. What time does the 3 P.M. bus get to Lincoln?

____

3. What is the earliest bus I can take to York on Sunday?

____

4. If I have to be in Lincoln by 4 P.M. on Monday, what time do I have to get the bus in York that day?

____

5. Now make up a question of your own about the bus schedule. Give your question to a friend to answer.

____

____

____

Name________________________ Date____________

Lesson Outline

Lesson 1

Minerals and Rocks

Fill in the blanks. Reading Skill: **Sequence of Events** - questions 7, 9, 11

What Are Minerals?

1. Rocks are made from materials called ______________.
2. A mineral is a substance found in nature that is not a plant or a(n) ______________.
3. An example of a rock that is made up of many different minerals is ______________.
4. An example of a rock that is made up of only one type of mineral is ______________.
5. You can tell one mineral from another by looking at color and ______________.
6. When you rub a mineral on something it leaves a mark called a(n) ______________.

How Are Rocks Formed?

7. A rock that forms when melted rock cools and hardens is called ______________ rock.
8. Melted rock that flows to the surface cools quickly and forms ______________.

Fill in the blanks.

9. When sand, mud, and pebbles pile up at the bottom of rivers, lakes, and oceans, _______________ rock is formed.

10. Examples of sedimentary rock include sandstone, shale, and _______________.

11. A rock that changes form through squeezing or heating is called _______________ rock.

12. Examples of metamorphic rock include gneiss and _______________.

13. There are four ways that large rocks break apart into tiny pieces. They are _______________, _______________, _______________, and _______________.

How Do We Use Rocks and Minerals?

14. Rocks are used to make roads, walls, and _______________.

15. Cement is made with crushed _______________.

16. Glass is made from tiny rocks pieces called _______________.

17. People cannot live without _______________, a mineral in food.

18. A mineral called _______________ is used to make pencils.

Name________________________ Date____________

Interpret Illustrations

Lesson 1

How Are Rocks Formed?

A diagram uses pictures and words to show a process. A diagram can also classify information. The diagram below shows kinds of rocks and how they are formed.

Use the diagram to answer the following questions.

1. What type of rock is granite? ________________
2. What type of rock is marble? ________________
3. What type of rock is siltstone? ________________
4. What rock is squeezed and heated to form marble? ________________

Name______________________________ Date_______________

Interpret Illustrations

Lesson 1

Rocks and Minerals

Rocks are everywhere you look. Study the pictures below. Think about how rocks are used in our everyday lives.

Use the diagram to answer the questions.

1. What metal is used to make pans? ________________
2. How is glass made?

   ____
3. What kind of rock is used to make statues? ________________
4. Look around you. Can you find things that are made of rock? Make a list.

____

Name______________________ Date__________

Lesson Vocabulary

Lesson 1

Minerals and Rocks

Fill in the blanks.

Vocabulary
minerals
metamorphic rock
igneous rock
sedimentary rock
chalk
scratch
property

1. Color is a(n) ______________ that is used to describe and identify rocks.
2. A kind of rock that is formed when melted rock cools and hardens is ______________.
3. One way to identify a mineral is to perform a(n) ______________ test.
4. A kind of rock formed when bits of sand, mud, and pebbles pile up and cement together is ______________.
5. A very soft rock that you can write with is called ______________.
6. Rocks are made from materials called ______________.
7. A kind of rock that has changed form through squeezing and heating is ______________.

Answer the question.

8. How do people use rocks?

____

____

____

Name ______________________________ Date ____________

Cloze Test
Lesson 1

Minerals and Rocks

Vocabulary

minerals	metamorphic	hard
marble	texture	streak

Fill in the blanks.

All rocks are made from types of materials called ________________. One way of identifying a mineral is to look at the color of its ________________. A mineral can also be identified by its color and ________________. Another property of rocks and minerals is how soft or ________________ they are.

There are three types of rocks: igneous, sedimentary, and ________________. An example of a type of rock that is made from squeezing and heating is ________________.

Name________________________ Date____________

Lesson Outline

Lesson 2

Kinds of Soils

Fill in the blanks.  Reading Skill: **Sequence of Events** - questions 3, 5, 8, 20, 21

What Makes Up Soil?

1. Soil is a mixture of small rocks, sand, minerals, and ______________.
2. A material that was once living or was formed by living things is called ______________.
3. The top layer of soil is called ______________.
4. Topsoil contains a lot of ______________ and minerals.
5. Below topsoil lies a layer of soil called ______________.
6. Subsoil holds water and some ______________.
7. There is no ______________ in subsoil.
8. Beneath subsoil is solid ______________.
9. Topsoil forms at Earth's ______________.
10. Rocks break into very small ______________ and mix with humus.

Fill in the blanks.

11. Different soils have different types of ______________ and minerals in them.

12. Some soils hold more ______________ than other soils.

13. A kind of soil with clay and sand is called ______________.

14. Soil layers can be thick or thin. Plants grow better in ______________.

15. Layers of soil on mountains are thin because ______________ pulls the soil downhill.

16. If soil is not taken care of, it can turn to ______________, which is dry, dusty soil.

Name________________________ Date____________

Layers of Soil

A diagram uses pictures and words to describe a thing or a process. This diagram gives you a close-up look at the layers of soil. The labels on the right side of the diagram name different layers of soil.

Use the diagram to answer the questions.

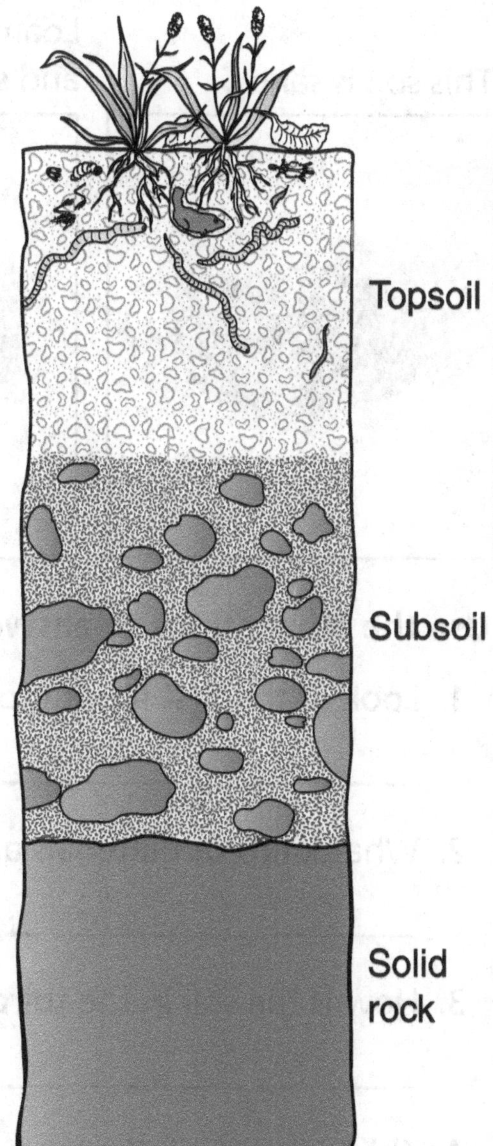

1. Where is topsoil found among the different layers of soil?

____

____

2. What lies directly beneath topsoil?

____

3. What lies directly beneath subsoil?

____

4. Compare the kinds of things you see in the different layers of soil.

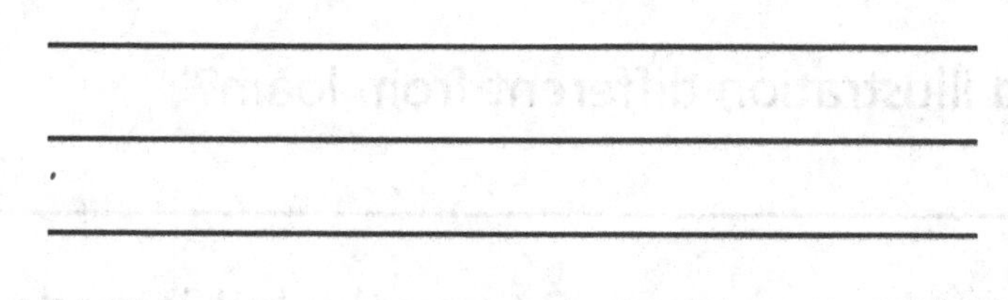

____

____

____

____

____

____

Name______________________________ Date______________

Interpret Illustrations
Lesson 2

What Makes Up Soil?

Soil is made up of many different materials. There are many kinds of soil. Study the illustrations below. Think about how soils are alike and different.

This soil is sandy.

Loam contains clay and sand.

This soil contains red clay.

Use the illustrations to answer the questions.

1. Look at the first illustration. What kind of soil is shown?

   ____

2. What kinds of materials are found in loam?

   ____

3. How is the soil in the third illustration different from loam?

   ____

4. Think about soil that you see in your town. Do you think it is made up of more sand, or more clay? Explain.

   ____

   ____

   ____

Name________________________________ Date______________

Lesson Vocabulary
Lesson 2

Kinds of Soils

Vocabulary

- plants
- topsoil
- soil
- rocks
- solid rock
- minerals
- humus
- loam

Fill in the blanks. You can use a word more than once.

1. A material that was once living or was formed by living things is ________________.
2. ________________ is made up of many different materials including small rocks, sand, minerals, and silt.
3. Soil is a mixture of small ________________, sand, minerals, and silt.
4. The top layer of soil is called ________________.
5. Subsoil lies between topsoil and ________________.
6. Subsoil does not have the decayed plant and humus found in ________________.
7. The layer of soil that holds water well is ________________.
8. Different soils have different kinds of rock and ________________ in them.
9. The type of soil in a particular place affects the kinds of ________________ that can grow there.
10. Plants grow well in ________________, a kind of soil with clay and sand.

Name ____________________ Date ____________

Cloze Test

Lesson 2

Kinds of Soils

Vocabulary

soil	air	subsoil	topsoil

Fill in the blanks. You can use a word more than once.

______________ is a mixture of humus, minerals, and small rock particles. Worms, insects, and roots growing in the soil create spaces for water and ______________. Usually, there are two layers of soil. The top layer, which has water and minerals for plant growth, is called ______________. Below the topsoil is a layer of soil called ______________. Beneath this is solid ______________.

Name________________________________ Date____________

Lesson Outline
Lesson 3

Fossils and Fuels

Fill in the blanks. Reading Skill: **Sequence of Events** - questions 2, 8, 9, 18, 19, 20

How Are Fossils Formed?

1. The imprint or remains of something that lived long ago is a(n) ________________.
2. A fossil begins to form when a plant or ________________ dies.
3. Most fossils are found in ________________ rocks.
4. Fossils are also found in hardened tree sap called ________________.
5. Tree sap is a sticky ________________ that can harden like glue.
6. Living things can make prints or marks in solid rock called ________________.
7. Fossils called molds are often left by ________________.
8. Shell molds form after shells are buried in ________________ or mud.
9. When ________________ seep into a mold, they can form a fossil called a cast.

Fill in the blanks.

What Do Fossils Tell Us About the Past?

10. Fossils tell us about ______________ and animals of the past.

11. We know about ______________ by studying their fossils.

12. Fossils also tell us how ______________ has changed.

13. An example of an animal that does not live anymore, but looked like an elephant is the ______________.

14. Some of Earth's oldest plants are ______________.

What are Fossil Fuels?

15. Cars, trucks, and planes need ______________ in order to run.

16. Examples of fossil fuels are gasoline, coal, and ______________.

17. Fossil fuels formed from the ______________ of plants and animals that lived long ago.

18. Millions of years ago, ______________ covered large parts of Earth's land.

19. Layers of decayed plants formed a soft material called ______________.

20. Over time, the peat was buried under layers of mud and sand and slowly changed into ______________.

Name____________________________ Date______________

Interpret Illustrations

Lesson 3

How Are Fossils Formed?

Illustrations can show us how things are alike and how they are different. The illustrations below show us a cast and a mold.

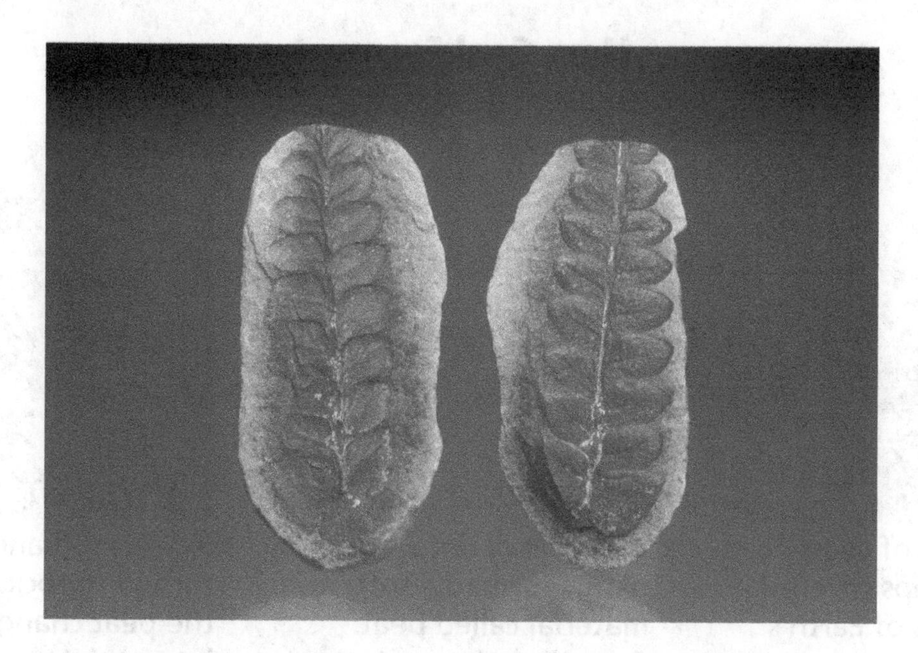

Answer these questions about the diagrams above.

1. What type of fossil is shown on the left? Explain how you know.

____

____

____

2. What type of fossil is shown on the right? Explain how you know.

____

____

____

What Are Fossil Fuels?

Diagrams can show the sequence in a process. The diagram below shows how coal was formed.

How Coal Forms

① Millions of years ago, swamps covered large parts of Earth's land. When swamp plants died, they sank to the bottom.

② Layers of decayed plants formed a soft material called peat. Over time the peat was buried under layers of mud and sand.

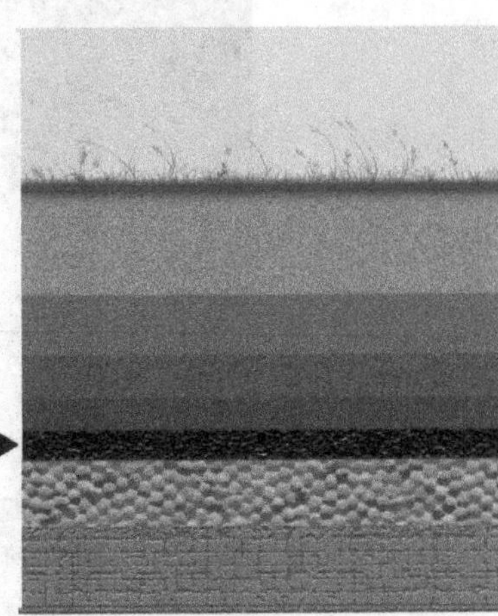

③ The mud and sand turned to rock. Slowly the peat changed into coal.

Answer these questions about the diagram above.

1. What happened to the swamp plants when they died?

____

2. What material formed from the layers of decayed plants?

____

3. Look at the third illustration. How was coal formed?

____

____

4. When was coal formed?

____

Name______________________ Date__________

Lesson Vocabulary
Lesson 3

Fossils and Fuels

Vocabulary
- fossil
- amber
- imprint
- mold
- cast
- fuel
- dinosaur
- mammoth

Fill in the blanks.

1. Scientists study the fossil bones of a(n) ________________ to build a model.
2. Hardened tree sap is called ________________.
3. An ancient relative of today's elephant might be the woolly ________________.
4. A living thing that pressed into materials long ago formed a shallow print called a(n) ____________.
5. A dinosaur bone is an example of the remains of a dinosaur. This bone is called a(n) ______________.
6. A kind of fossil that is an empty space in rock where something once was is called a(n) ________________.
7. When minerals seep into an empty space left in rock, a(n) ________________ forms.
8. A material that is burned for its energy is a(n) ________________.

Answer the question.

9. How do fossils tell us how Earth has changed?

____

____

____

Name ______________________________ Date ______________

Cloze Test

Lesson 3

Fossils and Fuels

Vocabulary

amber	**mold**	**sedimentary**
fossil	**imprints**	**cast**

Fill in the blanks.

A(n) ________________ is the imprint or remains of something that lived long ago. Most fossils are found in ________________ rock. Fossils are also found in ________________, which is hardened tree sap.

Shallow prints or marks in solid rock are called ________________.

A(n) ________________ is an empty space in rock where something once was. A(n) ________________ is a fossil made inside a mold.

Name______________________________ Date______________

Lesson Outline
Lesson 4

Water in Sea, Land, and Sky

Fill in the blanks. Reading Skill: **Sequence of Events** - questions 6, 8, 12 13, 15, 16

How Much of Earth's Surface is Water?

1. A little more than __________________ of Earth's surface is land.
2. Most of Earth's water is found in __________________.
3. Fresh water is found in ponds, streams, rivers, and most __________________.
4. Much of Earth's fresh water is trapped in large sheets of __________________.
5. The movement of water from place to place and from one form to another form is called the __________________.
6. When the Sun heats liquid water, the water __________________.
7. When water is in its gas form, it is called __________________.
8. When water cools into water droplets, it __________________.
9. Tiny droplets formed by condensed water vapor form __________________.
10. When clouds are heavy enough, water falls to Earth's surface in the form of rain, snow, sleet, or __________________.

Fill in the blanks.

How Can People in Dry Places Get Water?

11. A large wall that holds back the water from a river is called a(n) ____________.
12. Water from the river that gathers into a kind of lake is called a(n) ____________.
13. People build pipes or ditches called ____________ to carry water to places where it is needed.

Can We Get Water from the Ground?

14. Water held in rocks and soil below the surface is called ____________.
15. After it rains, water seeps through ____________.
16. After water seeps through soil, groundwater fills spaces in ____________ and ____________.

Why Should We Conserve Water?

17. When we conserve water, we use it wisely, not ____________.
18. You can conserve water by turning off water ____________ when you are not using them.

Name______________________________ Date____________

Interpret Illustrations

Lesson 4

How Much of Earth's Surface is Water?

Diagrams can use numbers to illustrate a concept or an idea. The diagram below shows how much of Earth's water is salt water and fresh water.

If All of Earth's Water Fit into 100 Cups

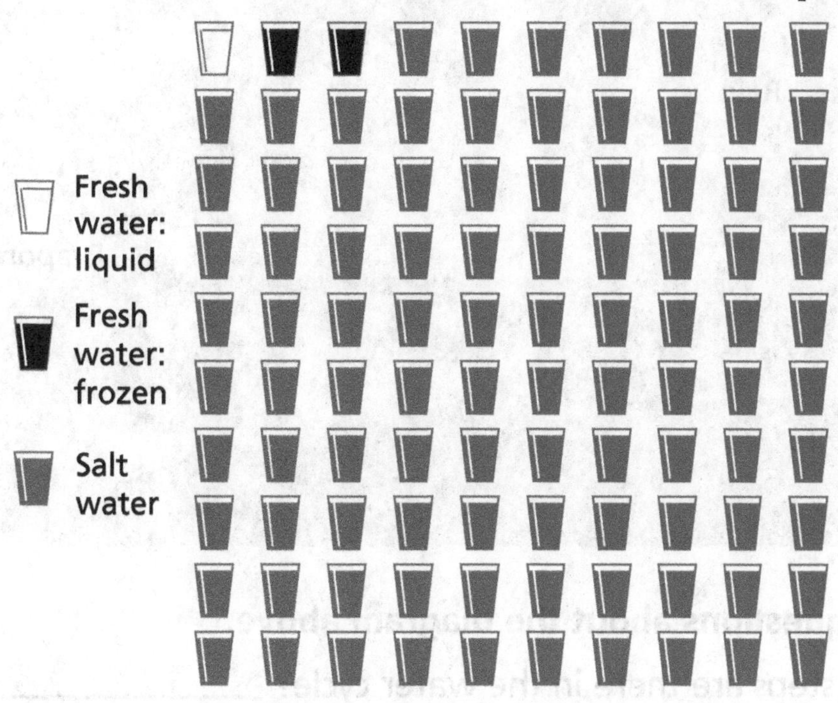

Answer these questions about the diagram above.

1. What is the title of this diagram?

____

2. What kind of water is found the most on Earth? How do you know?

____

3. Is there more fresh frozen water, or fresh liquid water? How do you know?

____

____

Name________________________ Date____________

Interpret Illustrations

Lesson 4

The Water Cycle

Diagrams can show the sequence in a process. The diagram below shows how water moves from place to place.

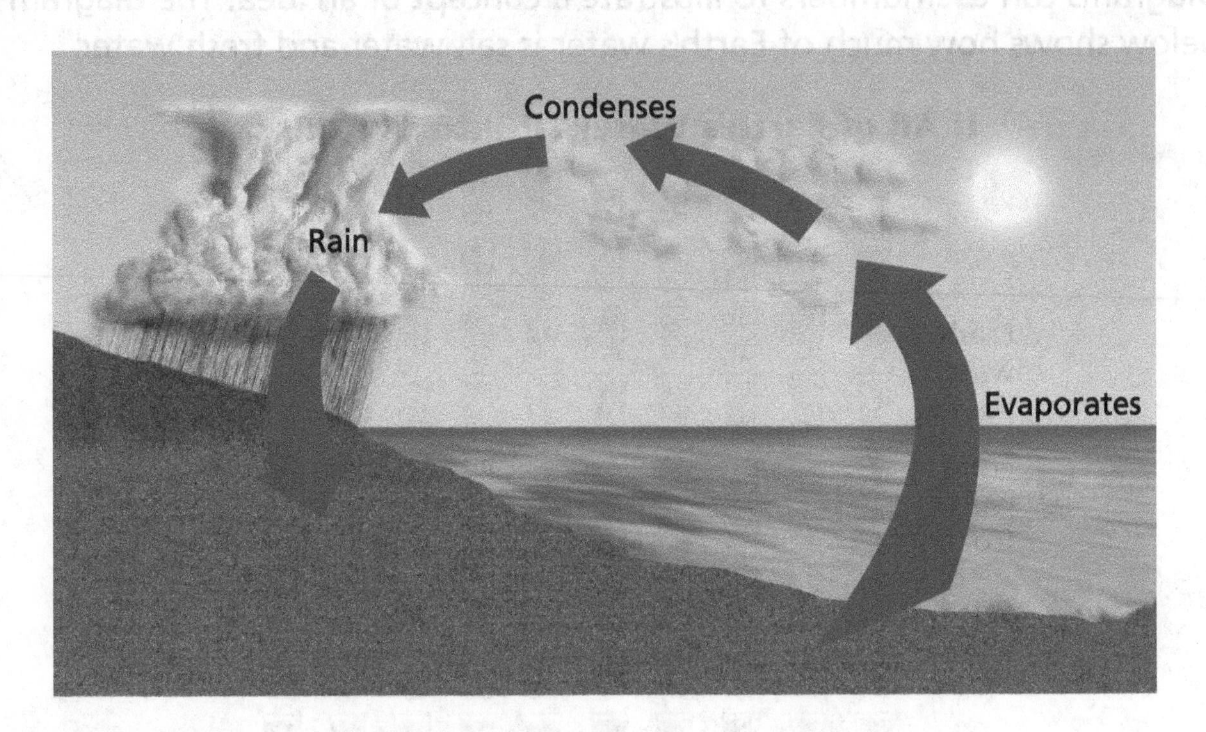

Answer these questions about the diagram above.

1. How many steps are there in the water cycle? ____________
2. What happens to water when it moves from the water into the air?

____

3. What happens to water after it condenses? ____________
4. What happens to water when it moves from clouds to the land?

____

Water in Sea, Land, and Sky

Vocabulary

water cycle
condenses
groundwater
conserve
reservoir
aqueduct
dam
evaporates

Fill in the blanks.

1. When water turns from a gas to a liquid, it ________________.
2. Water that seeps into the ground through cracks in rocks and soil is called ________________.
3. When water held back by a dam collects into a kind of lake, it forms a(n) ________________.
4. A pipe or a ditch that carries water is called a(n) ________________.
5. A large wall that is built across a river is called a(n) ________________.
6. When you use water wisely and not wastefully, you ________________ water.
7. Water travels from land to the air, from air to clouds, and then back to land again through the ________________.
8. When water changes from a liquid to a gas, it ________________.

Answer the question.

9. What is one way you can conserve water?

____

____

____

Name ______________________ Date ____________

Cloze Test
Lesson 4

Water in Sea, Land, and Sky

Vocabulary

water vapor	water cycle	evaporates	condenses
Sun	water	clouds	

Fill in the blanks.

The ________________ is the movement of water from place to place and from one form to another form. The ________________ powers the water cycle. When it heats liquid water, the water ________________. The ________________ rises through the air. When it cools, it ________________ into water droplets. These water droplets form ________________. When clouds are heavy enough, ________________ falls to Earth's surface.

Name______________________ Date____________

Lesson Outline

Lesson 5

Saving Our Resources

Fill in the blanks.  Reading Skill: **Sequence of Events** - questions 12, 13, 14

What Is a Natural Resource?

1. A material on Earth that is necessary or useful to people is a(n) ______________.
2. An important natural resource that comes from Earth's surface is ______________.
3. We use plants and animals for food, ______________ and many other things.
4. Fuels such as coal and ______________ come from below Earth's surface.
5. Iron, copper, and salt are ______________ that are also found below Earth's surface.

What Is a Renewable Resource?

6. A resource that can be replaced or used over and over again is called a(n) ______________.
7. Four kinds of renewable resources are soil, ______________, ______________, and ______________.

What Is a Nonrenewable Resource?

8. A resource than cannot be reused or easily replaced is a(n) ______________.

Name______________________ Date__________

Lesson Outline

Lesson 5

Fill in the blanks.

9. Nonrenewable resources include minerals, such as gems and ______________.

10. Coal, oil, and natural gas are examples of ______________, which are also nonrenewable resources.

What Is Pollution?

11. When harmful materials get into water, air, or land, ______________ occurs.

12. People cause pollution by adding soap and ______________ to water.

13. Harmful materials from cars, airplanes, and ______________ cause air pollution in many cities.

14. Land pollution comes from ______________ that takes up valuable space.

How Can You Conserve Resources?

15. A way to conserve resources is to ______________ our use of them.

16. When you use things again and again, you ______________ them.

17. We can recycle things such as glass, ______________, ______________, and ______________.

Name________________________________ Date______________

Interpret Illustrations

Lesson 5

What Is a Renewable Resource?

The illustrations below show ways people can use renewable resources.

Treated properly, the soil will support crops year after year.

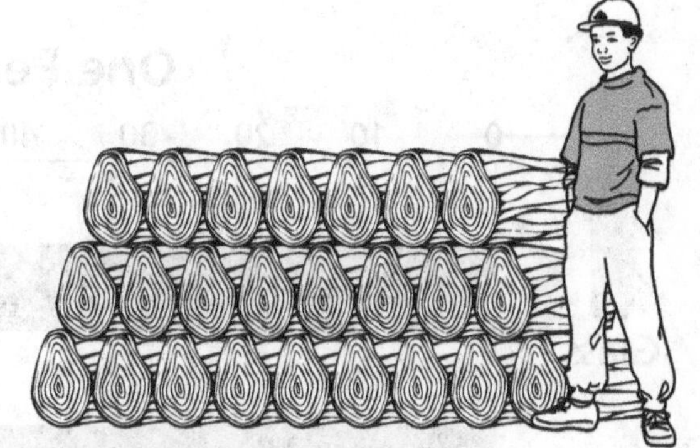

Like other plants, trees are a renewable resource. New trees can be planted to replace the ones we use.

Answer these questions about the illustrations above.

1. What crop is growing in the first illustration? How do people use this crop? ________________________________

2. What can be done to the soil so it can support crops year after year?

____

3. What renewable resource do the logs in the second illustration come from?________________________________

4. How can we renew our supply of trees?

____

Name ______________________ Date ____________

Interpret Illustrations
Lesson 5

How Can You Conserve Resources?

Charts and graphs often use numbers to show information. The numbers along the top of the graph tell how many. The names along the side of the graph tell what objects are counted.

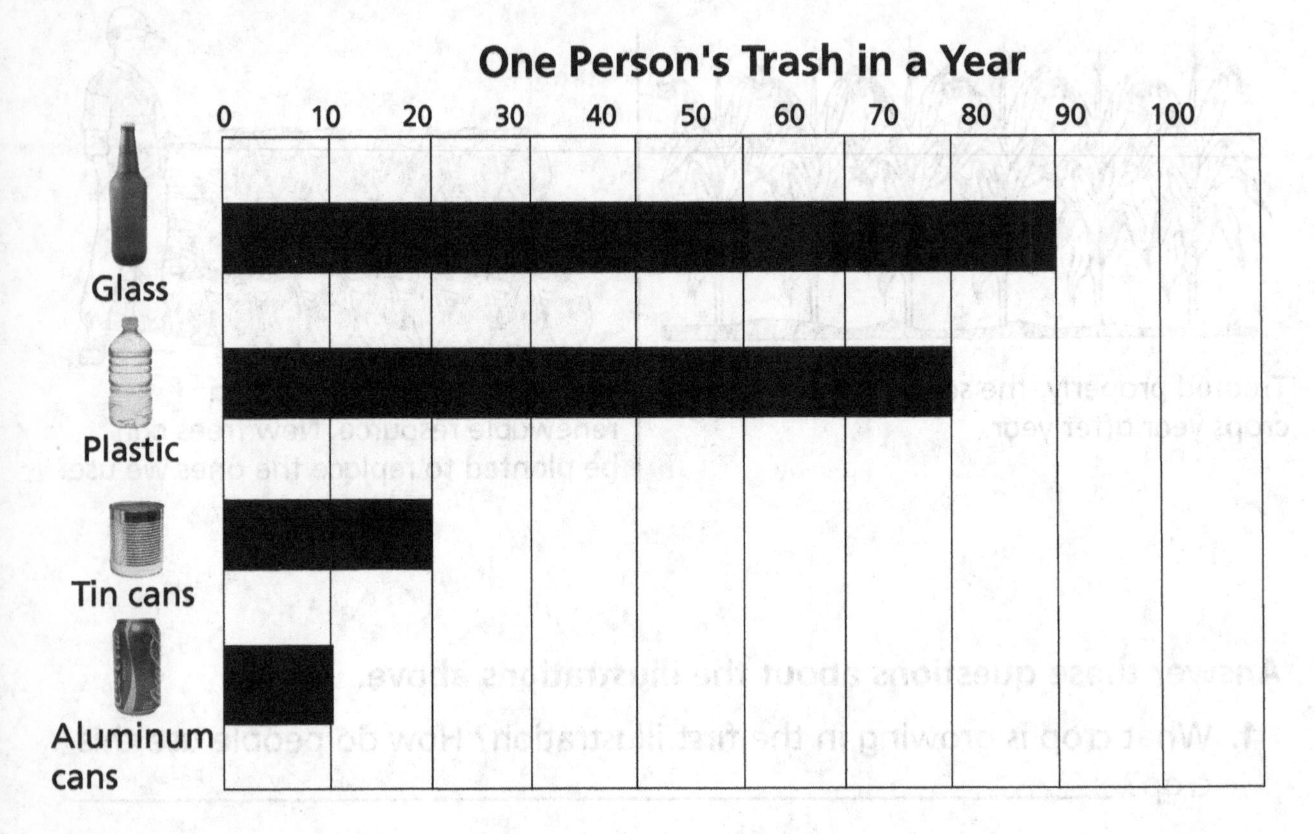

Answer these questions about the graph above.

1. What does this chart show? ______________________
2. How many pounds of tin cans does each person produce? __________
3. What items create the most trash? ______________________
4. What is one way you can conserve the resources you see in the chart?

____

Name________________________________ Date______________

Lesson Vocabulary
Lesson 5

Saving Our Resources

Fill in the blanks.

Vocabulary
- natural resources
- renewable resources
- nonrenewable resources
- pollution
- fossil fuels
- fertilizers
- recycle

1. ________________ are materials on Earth that are necessary or useful to people.
2. One source of water pollution comes from ________________ that people add to the water.
3. Minerals and metals are examples of ______________________.
4. One way to conserve a resource is to ____________, or treat something so that it can be used again.
5. Plants, animals, soil, and water are examples of ______________.
6. Resources that takes billions of years to form are the ___________.
7. When harmful materials get into water, air, or land, ________________ occurs.

Answer the question.

8. What is one way you can conserve resources?

____

____

____

Name________________________ Date____________

Cloze Test

Lesson 5

Saving Our Resources

Vocabulary

air pollution	land pollution	water pollution
recycle	nonrenewable	

Fill in the blanks.

When too much sand or soil settles into a lake, it can cause ________________. Earth's growing population can cause more trash and ________________. One way to reduce the amount of trash is to ________________ plastics, glass, and metals. Metals and minerals are examples of ________________ resources. Polluted rain water that harms trees and buildings comes from ________________.

Earth's Resources

Circle the letter of the best answer.

1. Rocks are made from substances found in nature called
- **a.** limestone.
- **b.** chalk.
- **c.** minerals.
- **d.** coal.

2. A type of rock that is formed from layers of sand, mud, and pebble is called
- **a.** sedimentary rock.
- **b.** metamorphic rock.
- **c.** igneous rock.
- **d.** all of the above.

3. You can compare minerals by looking at their
- **a.** color.
- **b.** hardness.
- **c.** texture.
- **d.** all of the above.

4. A material found in soil that was once living or was formed by a living thing is
- **a.** topsoil.
- **b.** humus.
- **c.** subsoil.
- **d.** minerals.

5. A kind of soil with sand and clay is
- **a.** humus.
- **b.** silt.
- **c.** subsoil.
- **d.** loam.

6. Fossils found inside a mold are known as
- **a.** casts.
- **b.** amber.
- **c.** imprints.
- **d.** silt.

7. A fuel is a material that is burned to get
   a. gasoline.
   b. oil.
   c. coal.
   d. energy.

8. Coal was formed millions of years ago from a soft material called
   a. silt.
   b. amber.
   c. limestone.
   d. peat.

9. When liquid water evaporates, it forms
   a. rain.
   b. water vapor.
   c. gas.
   d. ice.

10. When a dam is built across a river, it helps to form a (n)
   a. reservoir.
   b. ditch.
   c. aqueduct.
   d. well.

11. When harmful materials get into water, air, or land, it can cause
   a. natural resources.
   b. pollution.
   c. renewable resources.
   d. evaporation.

12. An example of a nonrenewable resource is
   a. a plant.
   b. a metal.
   c. an animal.
   d. soil.

Chapter Summary

1. What is the name of the chapter you just finished reading?

2. What are two vocabulary words you learned in the chapter? Write a definition for each.

3. What are two main ideas that you learned in this chapter?

Name______________________________ Date______________

Chapter Graphic Organizer
Chapter 6

Forces Shape the Land

Look at each word in the box. Decide if the change is a fast change or a slow change. Write the word in the correct word web.

tornado	plants	flood	wind	temperature
earthquake	water	volcano	chemicals	

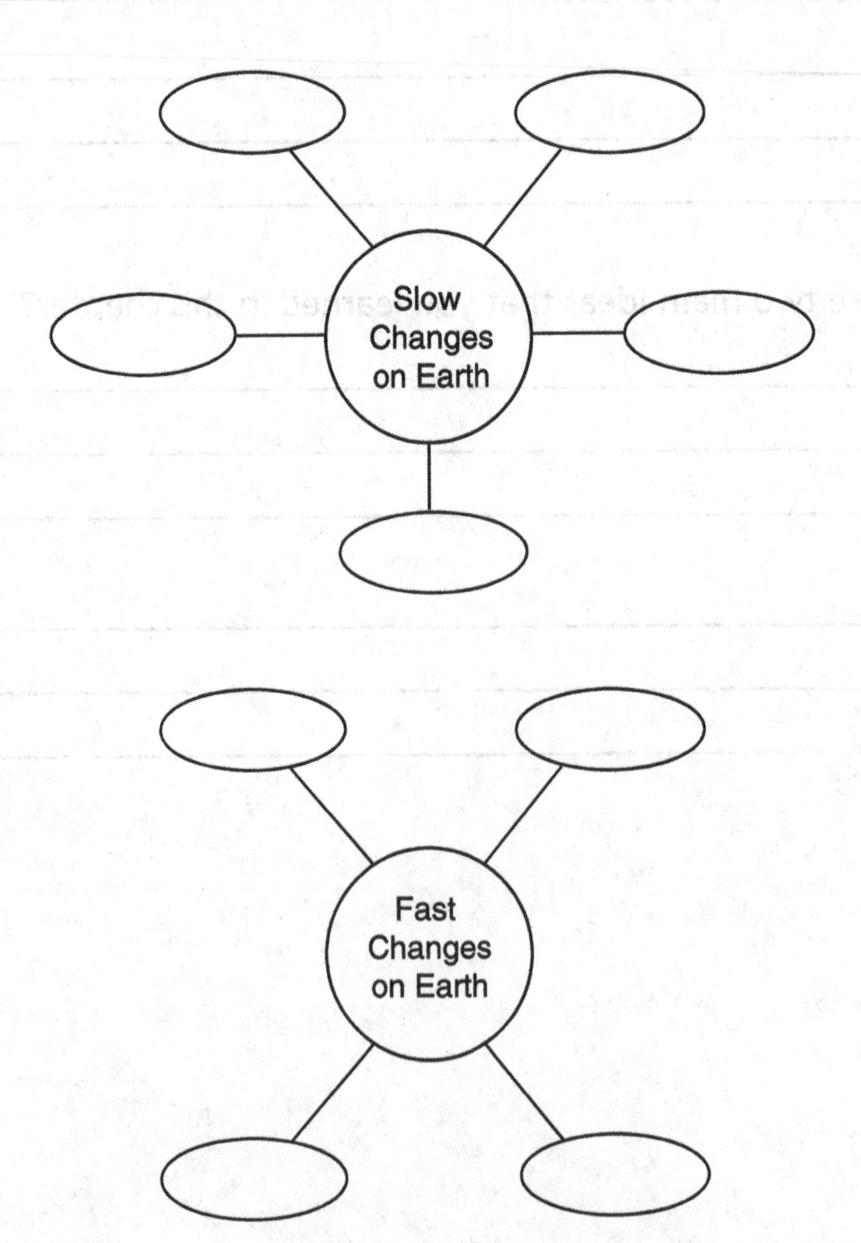

Name_________________________ Date_____________

Chapter Reading Skill

Chapter 6

Causes and Effects

This chapter in your textbook is about our constantly changing Earth. Since something has to make those changes happen, it's a good place to look for cause-and-effect examples! Use the chapter text to help you complete this cause-and-effect chart.

Cause	Effect
	Rivers overflow their banks.
Changing temperatures weather rocks.	
	A cave is formed.
weathering	
erosion	
	Ice moves weathered soil and rock.
hurricanes	
	Breaking rock makes the ground shake.
landslide	

Name________________________________ Date______________

Chapter Reading Skill
Chapter 6

Finding Causes and Effects

Read each example below. Circle each possible cause or effect.

Cause	Effect
1. Chris can't come to my birthday party **because** **a.** he is sick. **b.** he's my friend. **c.** it's on his dad's birthday! **d.** I didn't invite him. **e.** he reads books. **f.** I'm not having a party.	**1.** I had to write a book report, **so** **a.** I got a book from the library. **b.** my friend and I made cookies. **c.** I read the book. **d.** I went to bed early. **e.** I asked Dad if I could use his computer. **f.** I wrote it!
2. Racquel was the star of our class play **because** **a.** she forgot her homework. **b.** she's a very good actress. **c.** the play was her idea. **d.** she was the best one at tryouts. **e.** she likes strawberry yogurt. **f.** I didn't try out for the part!	**2.** We really liked our teacher this year, **so** **a.** I'm glad he's leaving. **b.** we gave him a surprise party. **c.** I took my dog for a walk. **d.** we had a class picture taken. **e.** we sang songs for him. **f.** we told the principal.

Name________________________________ Date______________

Lesson Outline

Lesson 6

Landforms

Fill in the blanks.  Reading Skill: **Sequence of Events** - questions 16 and 17

What Are the Features of Earth's Surface?

1. A feature of Earth's surface is a(n) ____________.
2. The highest landform on Earth's surface is a(n) ____________.
3. Landforms that are shorter and rounder than mountains are called ____________.
4. A flat land with steep sides is called a(n) ____________.
5. Flat-topped hills or mountains are known as ____________.
6. The low land between hills or mountains is called a(n) ____________.
7. A deep, narrow valley with steep sides is a(n) ____________.
8. Mounds of wind-blown sand are called ____________.
9. Wide, flat lands are called ____________.
10. Large streams of water that flow across the land are ____________.

Name ______________________ Date ____________

Lesson Outline

Lesson 6

Fill in the blanks.

11. A body of water with land all around it is a(n) ____________.

12. Very large bodies of salt water that cover three-fourths of Earth's surface are ____________.

13. The part of Earth where the ocean meets the land is the ____________.

14. A part of a lake or ocean that extends into the land is called a(n) ____________.

15. Huge masses of ice that move slowly across the land are called ____________.

What Is Earth's Surface Like in the United States?

16. Mountains that run from north to south and cover much of the west are called the ____________.

17. The Great Plains, where much of the food we eat in the United States is raised, are between North Dakota and ____________.

18. The ____________ are among Earth's largest lakes.

19. The Mississippi River empties into the ____________.

Name________________________ Date____________

What Are the Features of Earth's Surface?

The illustration below shows some of the landforms found on Earth.

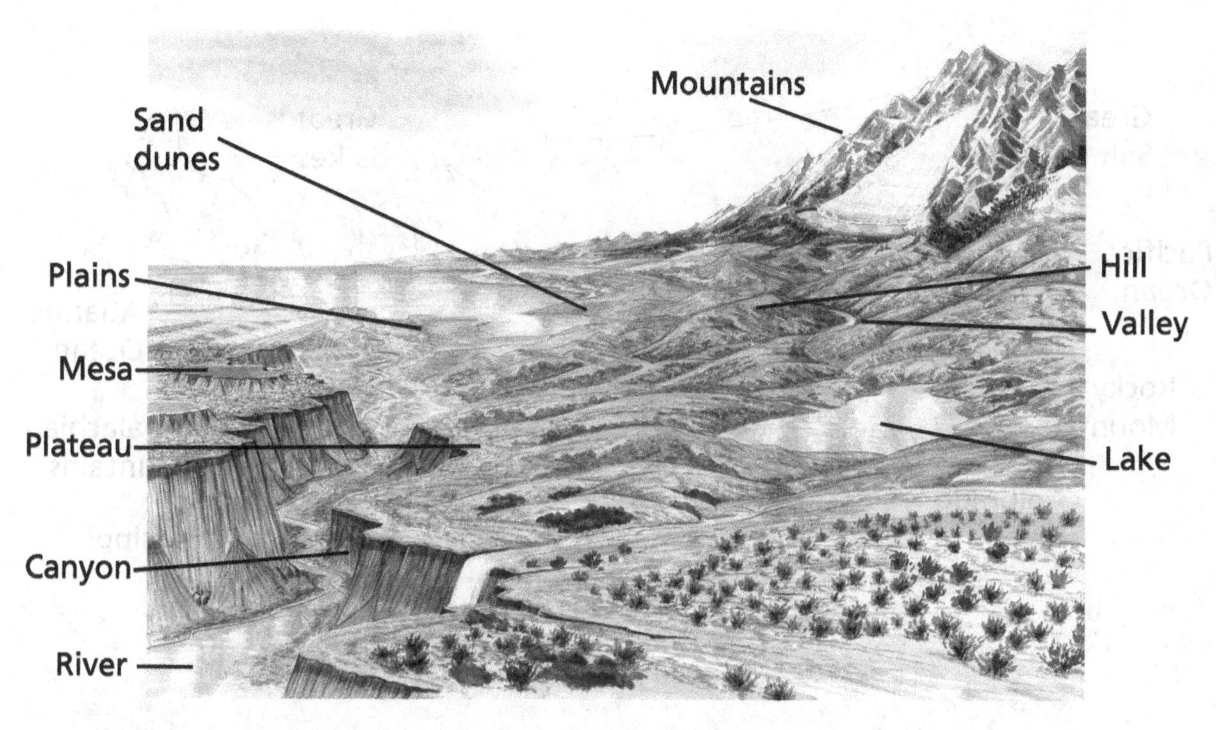

Answer these questions about the illustration above.

1. Which landform has flat land with steep sides? ____________
2. Which landform is the low land between hills or mountains?

____

3. What two bodies of water are shown in the diagram?

____

4. Which of the landforms often has rivers flow along the bottom?

____

Name______________________________ Date______________

Interpret Illustrations

Lesson 6

What Is Earth's Surface Like In the United States?

The map below is a topograhical map of the United States. This kind of map shows landforms.

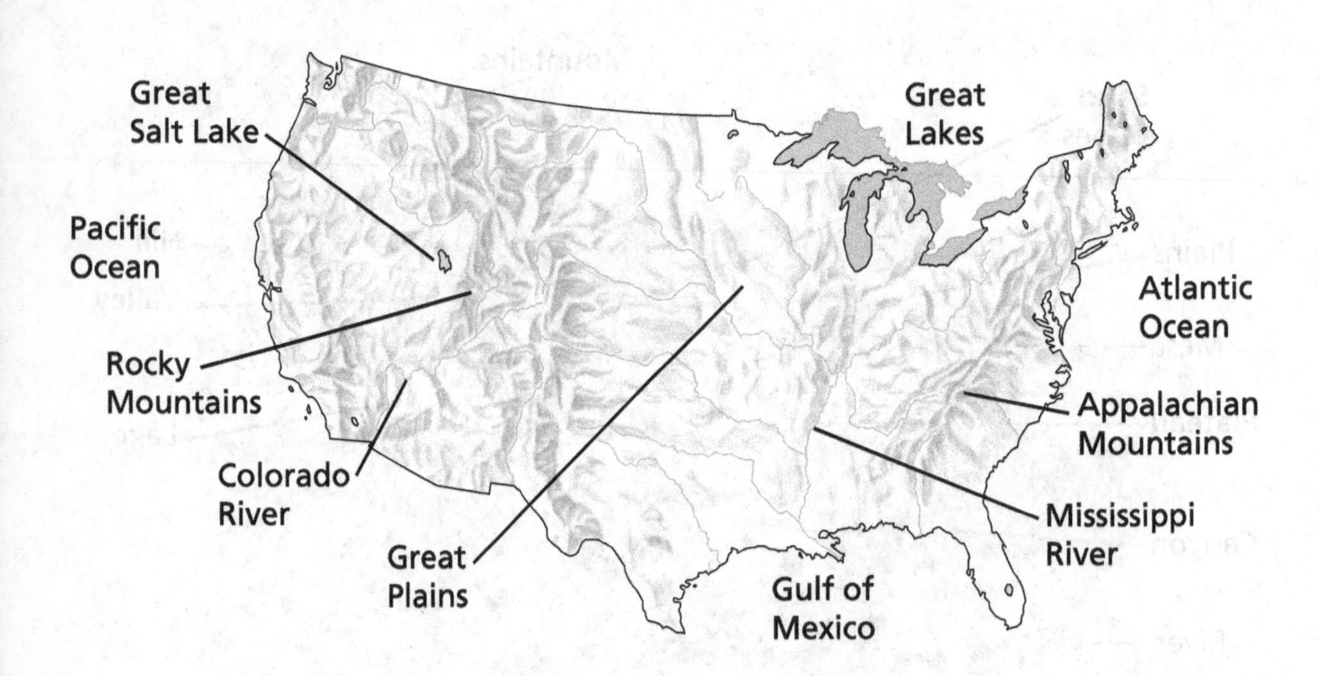

Answer these questions about the map above.

1. Which mountain range covers much of the west?

____

2. Which river runs down the middle of the United States?

____

3. The Mississippi River empties into the ______________________.

4. What is the flat land in the center of the United States?

____

Name______________________________ Date______________

Lesson Vocabulary
Lesson 6

Landforms

Match the letter of the vocabulary word to its definition.

Vocabulary

a. landform
b. mountain
c. valley
d. sand dunes
e. plain
f. rivers
g. coast
h. mesas
i. glaciers
j. canyon

_____ **1.** a feature of Earth's surface

_____ **2.** the low land between hills or mountains

_____ **3.** large streams of water that flow across land

_____ **4.** the highest landform with pointed tops

_____ **5.** flat-topped hills or mountains

_____ **6.** a deep, narrow valley with steep sides

_____ **7.** mounds of sand

_____ **8.** wide, flat lands

_____ **9.** huge masses of ice that move slowly across land

_____ **10.** where the ocean meets the land

Name______________________________ Date______________

Cloze Test
Lesson 6

Landforms

Vocabulary

hills	plains	lake
rivers	bay	landform

Fill in the blanks.

A(n) ________________ is a feature of Earth's surface. Landforms that are shorter and rounder than mountains are called ________________. Wide, flat lands are called ________________. Different bodies of water cover large parts of Earth, too. A body of water that is surrounded by land is called a(n) ________________. A body of water that extends into the land is called a(n) ________________. ________________ are large streams of water that flow across the land.

Name________________________________ Date______________

Lesson Outline
Lesson 7

Slow Changes on Land

Fill in the blanks.  Reading Skill: **Sequence of Events** - questions 1-7, 9, 13, 14, 18

How Do Rocks Change?

1. The process that causes rocks on Earth's surface to crumble, crack, and break is ________________.
2. Weathering usually happens ________________.
3. Rushing water causes the weathering of rocks on beaches and ________________.
4. Rocks can be worn down by ________________ carrying sand.
5. Rocks are broken apart by frozen water and ________________ that get into spaces in the rocks.
6. Chemicals cause the ________________ in rocks to change.
7. When the minerals in rocks change, the rocks ________________.
8. The first step in the formation of a cave is water entering cracks in ________________.
9. A small opening forms when ________________ in the water soften minerals in the rock.
10. When the space in rock gets larger, a(n) ________________ has formed.

Name ______________________ Date __________

Lesson Outline
Lesson 7

Fill in the blanks.

What Happens to Weathered Rocks?

11. Weathered materials are moved around by ______________.
12. Like weathering, erosion is often a ______________ process.
13. Weathered materials are carried great distances by ______________ and streams.
14. Weathered materials are pulled down hills and mountains by ______________.
15. Strong winds blow ______________ and sand.
16. A huge mass of moving ice is a(n) ______________.
17. As glaciers move over land, they pick up and move ______________ and other things.

How Do People Change Earth's Surface?

18. To make more room for houses, people cut down ______________.
19. Ponds and swamps called ______________ are drained.

Name________________________________ Date______________

Slow Changes on Land

A diagram uses pictures and words to describe a thing or a process. Each part of this diagram shows a step in the process by which a cave forms. The parts are arranged in order. The labels describe what is happening in the drawings.

How a Cave Forms

Water runs through cracks in limestone.

Chemicals in the water soften the limestone. Water washes away the weathered rock. A small opening forms.

The process continues. A cave has formed.

Use the diagrams to answer the questions.

1. What is the title of the chart?

____

2. What type of rock does water run through in the first step of this process? ______________________________

3. What causes a small opening to form in the second part of this process?

____

4. How does this space lead to the formation of a cave?

____

____

5. Could a cave form if cracks were not in the limestone shown in the first part of the diagram? Explain your answer.

____

____

Name____________________________ Date______________

Interpret Illustrations

Lesson 7

What Happens to Weathered Rocks?

You can learn a lot from illustrations. Captions often tell about illustrations.

Glaciers form when more snow falls than melts away. When glaciers become large enough, they creep downhill.

Answer these questions about the illustrations above.

1. How does a glacier form?

____

2. What happens when a glacier becomes very large?

____

3. What force helps move rocks down the side of a mountain?

____

4. How is gravity helped in moving rocks down a mountain?

____

Name ______________________________ Date ____________

Lesson Vocabulary
Lesson 7

Slow Changes on Land

Fill in the blanks.

Vocabulary
rushing water
gravity
erosion
crumble
slowly
freezes
glacier
weathering
rocks
cave

1. The process that causes rocks on Earth's surface to crumble, crack, and break is ______________.
2. Weathering is usually a process that takes place very ______________.
3. On beaches and river bottoms, ______________ causes the weathering of rocks.
4. When rocks are exposed to chemicals, the minerals in the rocks ______________.
5. Weathered materials are moved from place to place by ______________.
6. Weathered materials are pulled down hills and mountains by ______________.
7. A huge mass of moving ice is a(n) ______________.
8. Wherever a glacier goes, it moves ______________ and other things in its path.
9. When water ______________ inside a rock, the rock may crack.
10. When water runs through cracks in limestone, a ______________ can form.

Name________________________ Date____________

Cloze Test

Lesson 7

Slow Changes on Land

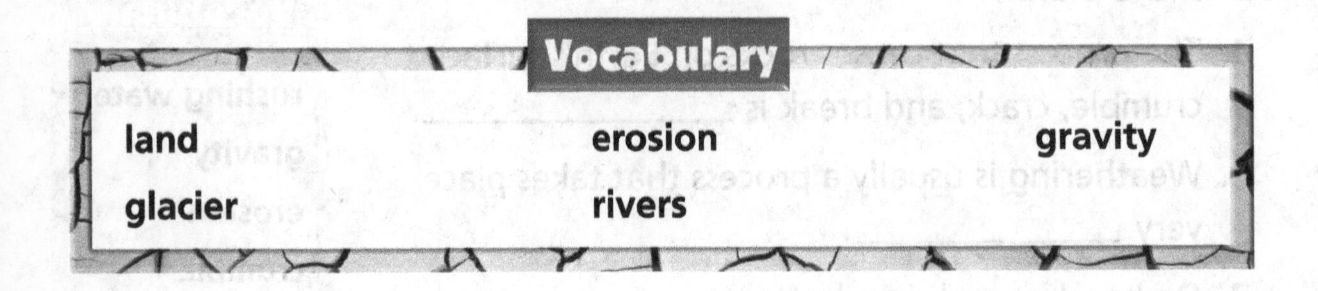

Fill in the blanks.

When weathered materials are carried away, ________________ occurs. These materials can be pulled down hills and mountains by ________________ or carried far distances by streams and ________________. A huge mass of moving ice is called a(n) ________________. Glaciers pick up rocks and other things as they move over ________________.

Name________________________ Date____________

Lesson Outline

Lesson 8

Fast Changes on Land

Fill in the blanks.  Reading Skill: **Cause and Effect** - questions 1, 2, 11, 12, 16, 19

How Can Land Change Quickly?

1. ____________, such as hurricanes, tornadoes, and floods, can change Earth's surface quickly.
2. A violent storm with strong winds and heavy rains is a(n) ____________.
3. Hurricanes are the largest and most powerful of all ____________.
4. Hurricanes form over ____________.
5. Hurricane winds move in a(n) ____________ pattern at high speeds.
6. Most hurricanes ____________ before they reach land.
7. A small, powerful windstorm over land is a(n) ____________.
8. A huge flow of water over dry land is a(n) ____________.
9. Floods carry away rocks and soil, and destroy ____________, ____________, and ____________.

How Do Earthquakes and Volcanoes Change the Land?

10. A sudden movement in the rocks that make up Earth's crust is a(n) ____________.
11. Forces within ____________ build up, then break Earth's rocks suddenly.
12. Earthquakes, heavy rains, or melting snow are some causes of ____________.
13. An opening in the surface of Earth is a(n) ____________.
14. Melted rock that flows from a volcano onto the surrounding ground is called ____________.

Name ______________________ Date ____________

Lesson Outline

Lesson 8

Fill in the blanks.

15. When a volcano erupts, ____________ and other materials flow onto Earth's surface.

16. The materials piling up around the opening cause a volcanic ____________ to form.

How Can Erosion Work Quickly?

17. At beaches and other sandy places, winds blow sand into sand ____________.

18. In the 1930s, dry weather in the Great Plains killed many ____________.

19. The dry soil was blown away with the wind because ____________ were not present to hold the soil in place.

20. During the 1930s, the Great Plains of the United States became known as the ____________ Bowl.

Name________________________________ Date________________

Interpret Illustrations

Lesson 8

How Can Land Change Quickly?

Hurricane damage

A tornado

____

2. What kind of damage was done by the hurricane?

____

3. Where do most tornadoes happen?

____

4. Do you think a tornado could happen in your state? Tell why or why not.

____

Name______________________ Date____________

Interpret Illustrations

Lesson 8

How a Volcano Forms

You can learn a lot from diagrams. Captions often tell about the diagram.

How a Volcano Forms

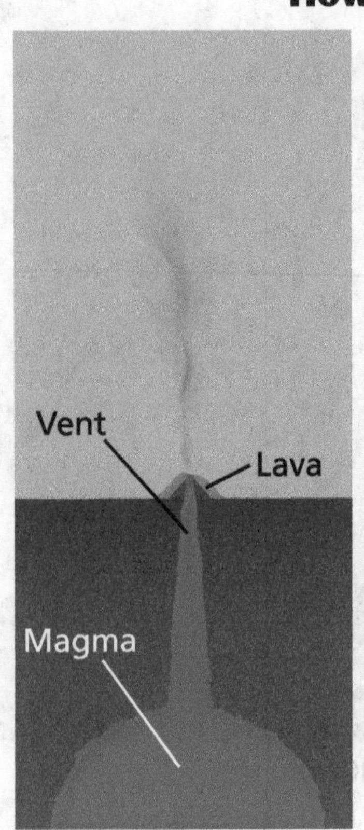

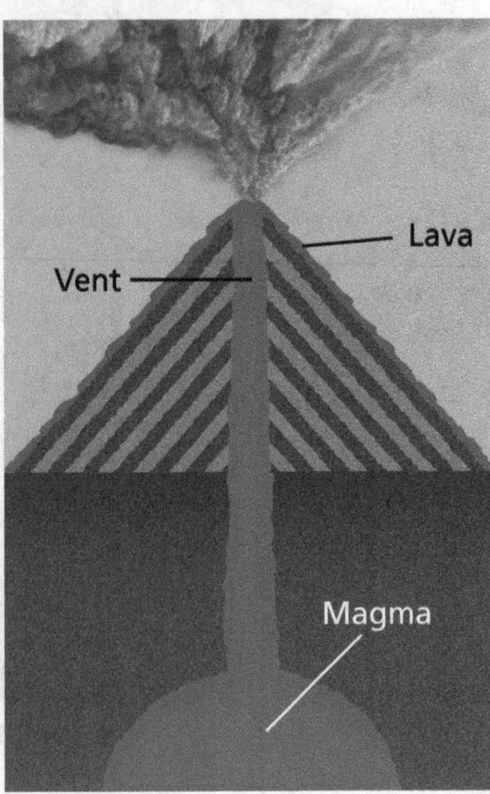

When a volcano erupts, lava and other materials flow onto Earth's surface. The materials pile up around the opening as they cool. Over time a mountain may form. Both the opening and the mountain around it are called a volcano.

Answer these questions about the diagram above.

1. What does this diagram show? ______________________
2. What flows onto Earth's surface when a volcano erupts?

   ____

3. What is lava called before it erupts from the volcano?__________
4. What happens as a result of a volcano?

   ____

Name________________________________ Date____________

Fast Changes on Land

Vocabulary

oceans
volcano
surface
floods
earthquake
hurricane
rock
landslide
lava
dust storms
die out
tornado

Fill in the blanks.

1. When weathering and erosion happen quickly, there are sudden changes to Earth's ______________.
2. A(n) __________________ is a violent storm with strong winds and heavy rains.
3. Most hurricanes form over the __________________.
4. Most hurricanes __________________ before they reach land.
5. Winds swirl very fast in a(n) __________________, sometimes 240 kilometers (approximately 150 miles) per hour or faster.
6. Heavy rains can cause __________________, as can melting snow or breaking dams.
7. A sudden movement of rock in Earth's crust is a(n) __________________.
8. Earthquakes, heavy rains, and melting snows can loosen soil, rocks, sand, and gravel, causing a(n) __________________ to occur.
9. An opening in the surface of Earth is called a(n) __________________.
10. Lava is melted __________________ that flows onto the ground.
11. During a volcanic eruption, __________________ and other materials flow out of a volcano onto the surface of Earth.
12. In the1930s, the Great Plains area of the United States were struck by terrible __________________.

Name ______________________ Date ______________

Cloze Test
Lesson 8

Fast Changes on Land

Vocabulary

gases	**volcano**	**fires**
mountain	**lava**	**damage**

Fill in the blanks. You may use a word more than once.

A(n) ________________ is an opening in the surface of Earth. Melted rock, ________________, pieces of rock, and dust are forced out of this opening. The word volcano is also a name of the ____________ that builds around this opening. Melted rock that flows out onto the ground is called ________________. Lava can cover everything in its path and start ________________. The ash, gases, and melted rock from volcanoes can cause a lot of ________________.

Name________________________________ Date____________

Forces Shape the Earth

Choose the letter of the best answer.

1. A feature of Earth's surface is a(n)
   a. erosion. b. landform.
   c. mineral. d. landscape.

2. When water runs through cracks in limestone, it forms a
   a. volcano. b. flood.
   c. cave. d. landslide.

3. All of the following are things on Earth that people change EXCEPT
   a. wetlands. b. earthquakes.
   c. forests. d. rocks.

4. A wide, flat land is called a
   a. beach. b. hill.
   c. plain. d. valley.

5. The highest landform that has steep sides and pointed tops is a
   a. mountain. b. plain.
   c. plateau. d. valley.

6. All of the following cause weathering in rocks EXCEPT
   a. chemicals. b. earthquakes.
   c. plant roots. d. temperature changes.

Name ______________________ Date ____________

Choose the letter of the best answer.

7. A huge mass of moving ice is
a. erosion. **b.** a glacier.
c. a landslide. **d.** a volcano.

8. Both weathering and erosion cause
a. changes in Earth's surface.
b. landforms to keep the same shape over time.
c. new types of rocks to form.
d. plants to grow.

9. A large and powerful storm is a(n)
a. earthquake. **b.** glacier.
c. hurricane. **d.** landslide.

10. An earthquake is a sudden movement in the rocks that
a. are located in a glacier. **b.** make up Earth's crust.
c. make up mountains. **d.** rest along the bottom of Earth's lakes, rivers, and oceans.

11. All of the following can cause landslides EXCEPT
a. duststorms. **b.** earthquakes.
c. heavy rains. **d.** melting snows.

12. Melted rock that flows out onto the ground during a volcanic eruption is called
a. dust. **b.** erosion.
c. lava. **d.** limestone.

Name ______________________ Date ____________

Crossword

Read each clue. Write the answer.

Across

4. Earth's necessary materials
7. Feature on Earth's surface
11. Mass of ice
13. Flows from some volcanoes
14. Flat area above surrounding land
15. Covers Earth but isn't water

Down

1. Frozen water
2. Land between two hills
3. Make dirty
5. Use again
6. Process that carriers away soil
8. Not plant or animal
9. Opening in Earth's surface
10. Treat and use again
12. Land with few hills

Name ______________________ Date ____________

Crack-a-Code

Code Key

A C E G I L N O P R S T U V Y

Use the Code Key to help you decode each word. Then draw a line to its meaning.

1. ______________________ a. a huge mass of ice

2. ______________________ b. flat land rising above the land around it

3. ______________________ c. opening in Earth's surface

4. ______________________ d. land between hills

5. ______________________ e. use something again

6. ______________________ f. land with few hills

7. ______________________ g. process that carries away rock and soil

Name________________________________ Date______________

Correct-a-Word

Word Box				
pollution	glacier	hurricane	mineral	reduce
recycle	fertilizer	plateau	weathering	volcano

One word in each sentence below is wrong. Cross it out and write the correct word above it. One has been done for you. Use the Word Box to check your spelling.

1. A flat land with steep sides is a ~~plate.~~ plateau
2. A glass is a huge mass of ice moving across land.
3. Strong winds blow during a hericane.
4. Inside a vulcan is melted rock that comes out as lava.
5. Remember to bicycle papers, cans, and plastic objects.
6. Wandering is a process that causes rocks to crumble.
7. If it's not a plant or animal, it must be a minerey.
8. Making clean water dirty is collision.
9. We use furtolizor in our garden to help the plants grow.
10. When you relose, you make less of something.

Name ____________________ Date ____________

Chapter Graphic Organizer
Chapter 7

Earth's Weather

Complete the chart.

Characteristics of Weather	How It Is Measured
temperature	
	barometer
wind speed	
	wind vane
precipitation	

Fill in the correct words to complete the water cycle chart.

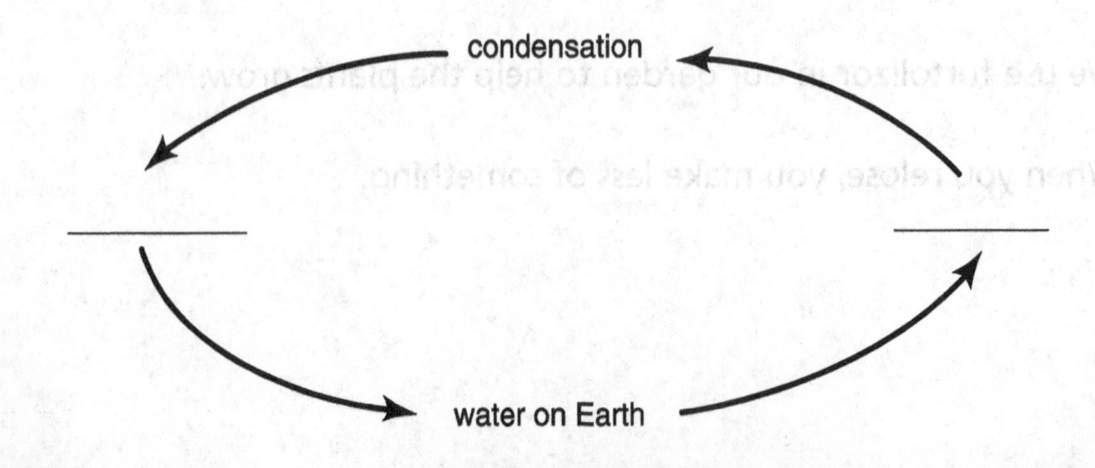

Name____________________________ Date______________

Summarize

When you **summarize**, you tell the important idea of a story or an article. A summary does not need to include the small details. It can be written in one to three sentences.

Read the article. Then write a summary in one or two sentences.

Blizzards

Have you ever been in a really bad snowstorm? If so, you might remember snow that was piled very deep. A blizzard is a name for a very strong and very cold wind that causes snow to blow up from the ground. Blizzards can be very dangerous. Weather conditions include freezing temperatures, high wind speeds, and great big snow drifts. The winds cause snow to blow around so much that people can't see very far. Blizzards can be very unsafe for people who are outside. Most blizzards happen in a cold air mass. In the Great Plains, some winter dust storms are called black blizzards.

My Summary

____

____

____

____

____

Name ______________________ Date ____________

Chapter Reading Skill
Chapter 7

Weather

You can find lots of information about the weather in the daily newspaper. You can find the weather all across the country and in other cities. You can find the details of your local weather. Find a local newspaper. Look up the weather in your area. **Draw a picture to show the weather, then write a short summary.**

My Picture

My Summary

____

____

____

____

____

Name__________________________ Date______________

Lesson Outline
Lesson 1

The Weather

Fill in the blanks. Reading Skill: **Summarize** - questions 4, 10, 11, 13

Where Is Air?

1. The air that surrounds Earth is called the ____________.
2. The atmosphere is made up of different gases and ____________.
3. Some of the dust in the atmosphere comes from fires and ____________ on Earth.
4. The four layers of the atmosphere are called the
   a. ____________,
   b. ____________,
   c. ____________, and the
   d. ____________.

What Is Air Temperature?

5. ____________ is a measure of how hot or cold something is.
6. Temperature is measured with a(n) ____________.

Fill in the blanks.

7. The Earth's North and South Poles do not get as much sunlight as places near the ____________________.

8. The Sun stays ____________________ in the sky near the Earth's poles.

What Is Air Pressure?

9. The pressing down force of the air on Earth is called ____________________.

10. You can feel air pressure when you climb up a(n) ____________________, ride in a(n) ____________________, or fly in a(n) ____________________.

What Happens When Air Moves?

11. Wind happens when air moves from an area of ____________________ pressure to an area of ____________________ pressure.

12. Large bodies of air are called ____________________.

13. The weather in the place where air masses meet may be cloudy, rainy, or ____________________.

Name________________________________ Date______________

Interpret Illustrations

Lesson 1

Where Is Air?

The diagram below shows the layers of the Earth's atmosphere.

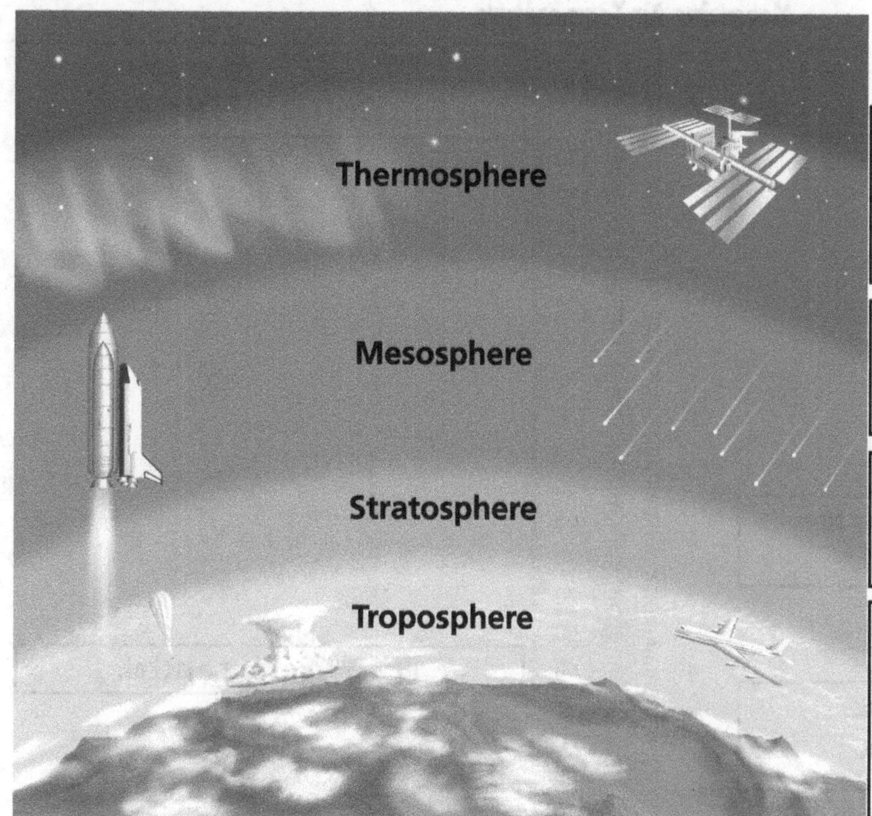

Layers of the atmosphere

The thermosphere is the top layer of the atmosphere. Light displays called the *northern lights* happen here.

The mesosphere is between the stratosphere and the thermosphere.

In the stratospere the air is not still. Very strong winds called the *jet stream* are located here.

The troposphere is the layer of the atmosphere closest to Earth. In this layer all clouds, rains, snow, and thunderstorms occur.

Answer these questions about the diagram above.

1. How many layers are in Earth's atmosphere? ________________
2. What are the strong winds found in the stratosphere called?

   ____
3. In which layer of the atmosphere do clouds, rain, snow, and thunderstorm occur? ________________
4. In which layer of the atmosphere might you find a space station?

   ____

Name ____________________ Date ____________

Interpret Illustrations

Lesson 1

What Is Air Temperature?

Diagrams can show information by using captions and labels.

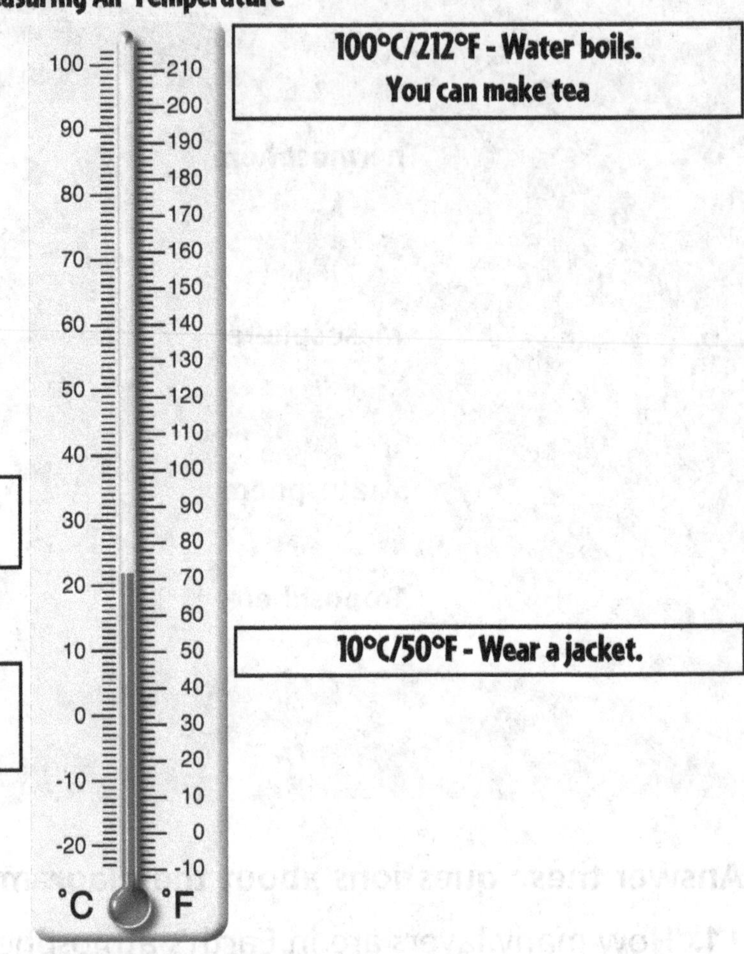

30°C/86°F - It's a good day to go to the beach.

0°C/32°F - Water freezes. It's cold, so snuggle up.

Answer these questions about the diagram above.

1. What is the title of this diagram? ____________________
2. At what temperature does water boil? ____________________
3. What would you wear to play outdoors if the temperature were 0°C?

____

____

4. What would you be able to do if it were 86°F? ____________________

____

Name ______________________ Date ____________

Lesson Vocabulary
Lesson 1

The Weather

Fill in the blanks.

Vocabulary
atmosphere
weather
air pressure
temperature
boils
freezes
wind
air masses

1. The air that surrounds Earth is where storms and other kinds of ______________________ take place.
2. The troposphere is one of the layers that makes up Earth's ______________________.
3. You would use a thermometer to measure ______________________.
4. When water reaches a temperature of 32°F, it ______________________.
5. When air moves from an area of high pressure to an area of lower pressure, the moving air is called ______________________.
6. When water reaches a temperature of 100°C, it ______________________.
7. Most weather happens where two bodies of air, called ______________________, meet.
8. The pressing down force of the air on Earth is called ______________________.

Answer the question.

9. How might you know when the air pressure around you changes?

____

Name____________________ Date__________

Cloze Test
Lesson 1

The Weather

Vocabulary

stratosphere	**mesosphere**	**atmosphere**
thermosphere	**troposphere**	**weather**

Fill in the blanks.

The air around Earth is made up of four layers, called the ____________________. The layer that is closest to the Earth's surface is the ____________________. All life on Earth exists here. This is where ____________________ takes place. Very strong winds called the jet stream are found in the ____________________. The layer that comes between the stratosphere and the thermosphere is the ____________________. The top layer of the atmosphere is called the ____________________.

Name ______________________________ Date ______________

Lesson Outline
Lesson 2

The Water Cycle

Fill in the blanks. Reading Skill: **Summarize** - questions 2, 3, 4, 9, 11

Where Does Water Go?

1. Everything around you is made up of ______________.
2. A form of matter that has a definite shape and takes up a definite amount of space is a(n) ________________.
3. A form of matter that takes up a definite amount of space, but does not have a definite shape is called a(n) ________________.
4. A form of matter that has no definite size or shape is called a(n) ________________.
5. When water changes from a liquid to a gas, the gas is called ________________.
6. Evaporation happens when a liquid changes into a(n) ____________.
7. When water vapor in the air ________________, it changes back into liquid water.
8. Condensation happens when a gas changes into a(n) ________________.

Name ______________________ Date ____________

Lesson Outline

Lesson 2

Fill in the blanks.

What Is the Water Cycle?

9. The never-ending path water takes between Earth and the atmosphere is called the ______________________.
10. Water that returns to Earth from the atmosphere is called ______________.
11. Precipitation can be in the form of
    a. ______________________,
    b. ______________________,
    c. ______________________, or
    d. ______________________.
12. The form of precipitation that falls to Earth depends mainly on the ______________.
13. Precipitation that soaks into the ground is called ______________________.

Name________________________________ Date______________

Interpret Illustrations

Lesson 2

Where Does Water Go?

Illustrations can teach us about concepts. Think about what you know about evaporation and condensation.

Where will the water in this lake go?

Where did this water come from?

Answer these questions about the diagram above.

1. What does the first picture show? ______________________________
____

2. What might cause the water in the lake to disappear?
____

3. What is on the spider's web? ____________________

4. What might cause the water to appear on the spider's web?
____

Name ______________________ Date ____________

Interpret Illustrations
Lesson 2

What Is the Water Cycle?

Diagrams can show what happens in a process by using labels and arrows.

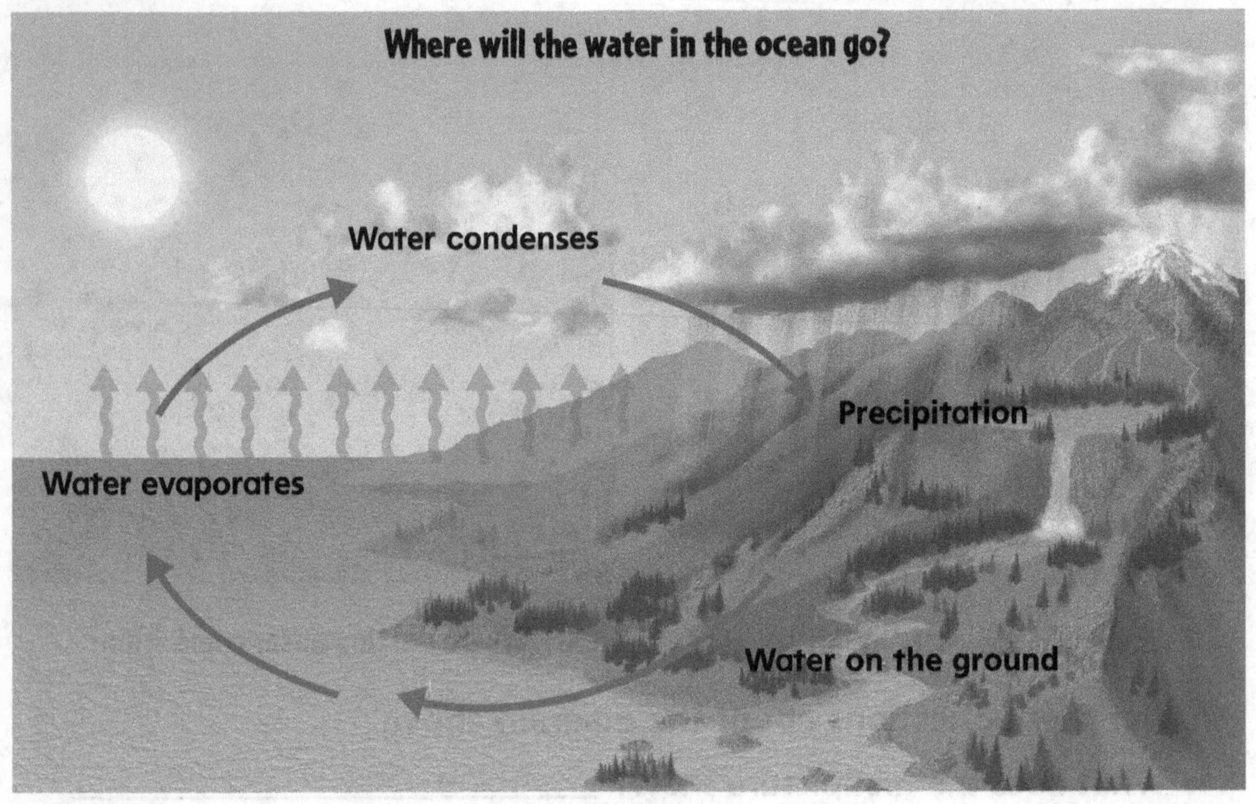

Answer these questions about the diagram above.

1. What happens when the Sun heats water in the ocean and on land?
   ____

2. What happens to the cooled water vapor in the air?
   ____

3. What is water that falls to Earth called? ____________________

4. What are lakes and rivers examples of? ____________________

Name ____________________ Date __________

Lesson Vocabulary

Lesson 2

Fill in the blanks.

Vocabulary
water vapor
evaporation
condensation
water cycle
precipitation
matter
groundwater

1. Rain, hail, sleet, and snow are all forms of ____________________.
2. The never-ending circle of water between Earth and the atmosphere is the ____________________.
3. When gas changes into a liquid, ____________________ takes place.
4. Water that changes into a gas is called ______________________________.
5. The changing of a liquid into a gas is called ____________________.
6. Solids, liquids, and gas are all forms of ____________________.
7. Precipitation that soaks into Earth after a rainfall is called ____________________.

Answer the question.

8. Why is water an important resource?

____

____

Name ________________________ Date ____________

Cloze Test
Lesson 2

The Water Cycle

Vocabulary

water vapor	condensation	precipitation
water cycle	evaporation	cools

Fill in the blanks.

The ________________ is the never-ending path water takes between Earth and the atmosphere. The Sun heats the water in the ocean and on land. The water changes from a liquid into ________________. The changing of the liquid into a gas is called ________________. When the water vapor in the air ________________, it changes back into liquid water. The change of a gas into a liquid is called ________________. When enough water has condensed in clouds, the water falls to Earth as ________________.

Name______________________ Date____________

Lesson Outline
Lesson 3

Describing Weather

Fill in the blanks. Reading Skill: **Summarize** - questions 1, 9

How Do You Describe Weather?

1. When scientists describe weather, they measure
   a. ______________________,
   b. ______________________,
   c. ______________________,
   d. ______________________.
2. The temperature of the air is measured with a(n) ______________________.
3. Air pressure is measured with a(n) ______________________.
4. How much precipitation has fallen is measured with a(n) ______________________.
5. Direction of the wind is measured by a(n) ______________________.
6. How fast the air is moving is measured with a(n) ______________________.

Name____________________________ Date______________

Fill in the blanks.

How Do You Read a Weather Map?

7. Scientists use symbols to show data on a(n) ______________________.

8. Scientists use the weather map to make weather

   ______________________.

9. The weather map on page D26 shows

   a. ____,

   b. ____, and other

   c. ____.

10. Some places, such as Phoenix, are having ____________________,

    ______________________ weather.

11. Other places, such as New York City, are ____________________ and

    ______________________.

12. Each color band represents an area with the same

    ______________________.

Name______________________________ Date______________

Interpret Illustrations
Lesson 3

How Do You Describe Weather?

You can learn a lot from illustrations and captions. Study the pictures.

A barometer measures air pressure.

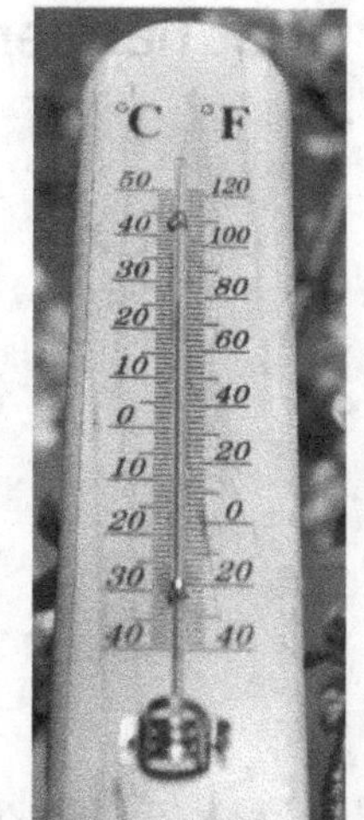

A thermometer measures the temperature of the air.

A rain gauge measures how much precipitation has fallen.

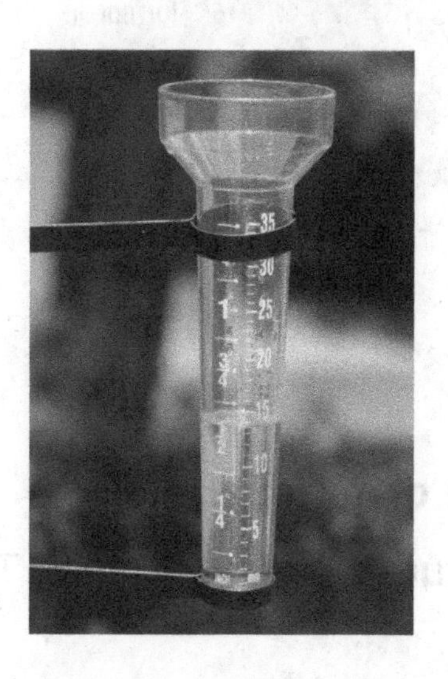

A weather vane indicates the direction of the wind. The arrowhead tells you where the wind is coming from.

Answer these questions about the diagram above.

1. Which instrument measures the direction of the wind?
   ____
2. What does a thermometer do? ______________________________
3. Which instrument measures air pressure? ____________________
4. Which instrument would you use to find out how much rain has fallen? ______________________________

Name________________________________ Date______________

Interpret Illustrations

Lesson 3

How Do You Read a Weather Map?

Diagrams can show what happens in a process by using captions and arrows.

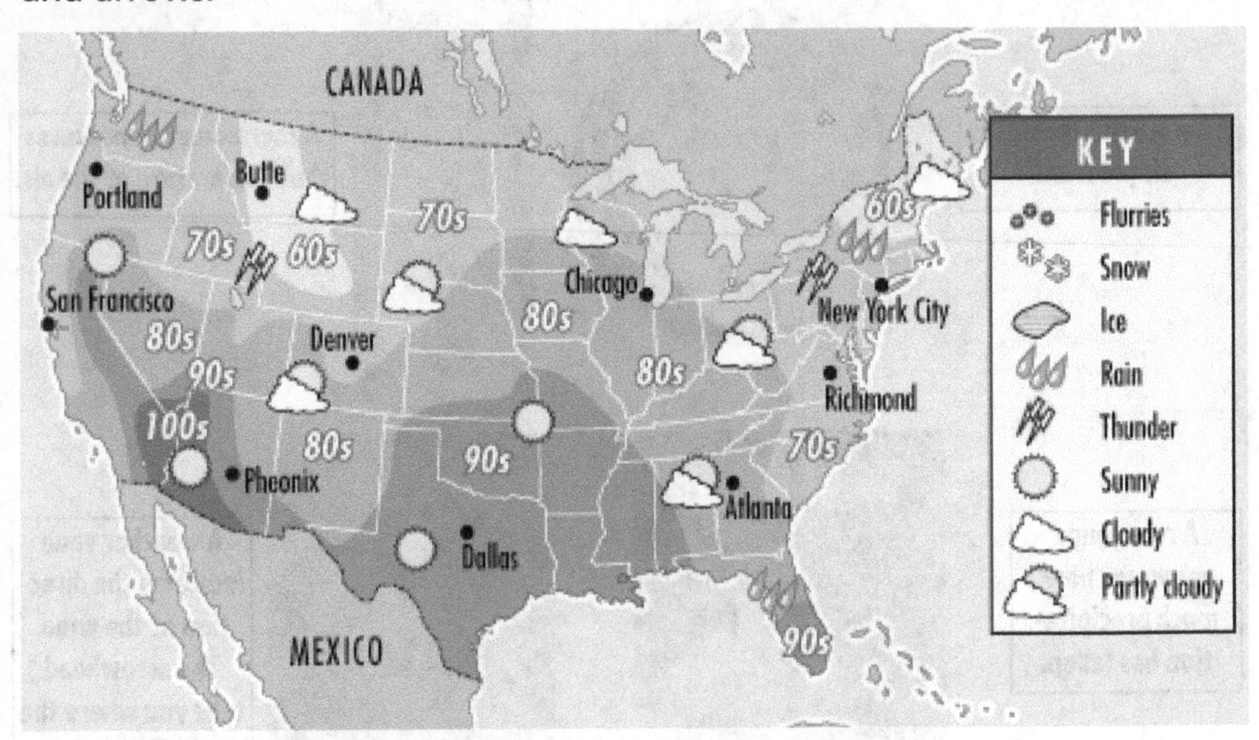

Answer these questions about the diagram above.

1. What is the predicted temperature range for Dallas, Texas?

____

2. What is the predicted cloud cover for San Francisco?

____

3. What is the lowest temperature range on the map?

____

4. What is one part of the United States that may have rain?

____

Name___________________________ Date____________

Describing Weather

Vocabulary

- **barometer**
- **thermometer**
- **precipitation**
- **direction**
- **speed**
- **symbols**
- **weather map**

Fill in the blanks.

1. An instrument that measures air pressure is called a(n) ____________________.
2. A weather vane is an instrument that measures wind ____________________.
3. A rain gauge is an instrument that measures ____________________.
4. Weather maps have ____________________ that show data.
5. An instrument that measures temperature is called a(n) ____________________.
6. You would use an anemometer to measure wind ____________________.
7. Weather conditions in the United States can be seen on a ________________________.

Answer the question.

8. How can a weather map help you know about weather?

____

____

Name ________________________ Date ____________

Cloze Test
Lesson 3

Describing Weather

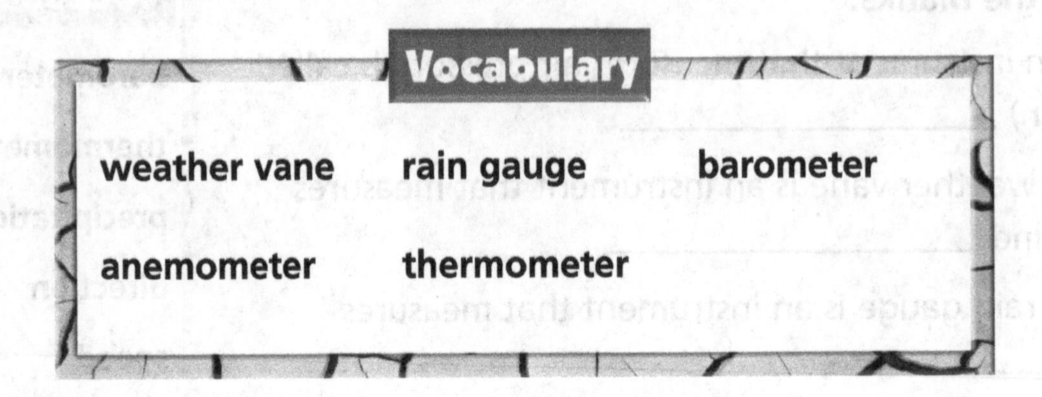

Fill in the blanks.

A ____________________ is an instrument you would use to measure air pressure. You would use a ____________________ to find out the temperature. The amount of precipitation can be found when you use a(n) ____________________. How fast the wind travels is measured with a(n) ____________________. A ____________________ indicates the direction of the wind.

Earth's Weather

Circle the letter of the best answer.

1. The air that surrounds Earth is called its

a. temperature. **b.** layers.
c. air pressure. **d.** atmosphere.

2. The layer of air that is closest to Earth is called the

a. troposphere. **b.** stratosphere.
c. mesosphere. **d.** thermosphere.

3. What the air is like at a given time and place is called

a. wind. **b.** weather.
c. air pressure. **d.** temperature.

4. Large bodies of air are called

a. wind. **b.** weather changes.
c. air masses. **d.** storms.

5. The changing of a liquid into a gas is called

a. groundwater. **b.** precipitation.
c. condensation. **d.** evaporation.

6. A form of matter that has a definite shape and takes up a definite amount of space is a

a. solid. **b.** gas.
c. liquid. **d.** none of the above.

7. The never-ending path of water between the atmosphere and Earth is called
   a. precipitation.
   b. condensation.
   c. the water cycle.
   d. matter.

8. When water changes from a liquid to gas, it is called
   a. water vapor.
   b. hail.
   c. sleet.
   d. snow.

9. A thermometer is used to measure
   a. air pressure.
   b. temperature.
   c. wind direction.
   d. wind speed.

10. Air pressure is measured with a(n)
   a. anemometer.
   b. weather vane.
   c. barometer.
   d. rain gauge.

11. Weather maps show all of the following EXCEPT
   a. current conditions.
   b. temperature ranges.
   c. cloud cover.
   d. condensation.

Chapter Summary

1. What is the name of the chapter you just finished reading?

____

____

2. What are two vocabulary words you learned in the chapter? Write a definition for each.

____

____

____

3. What are two main ideas that you learned in this chapter?

____

____

____

____

____

____

Name________________________ Date____________

Chapter Graphic Organizer
Chapter 8

Earth and Space

Look at each word in the box. Decide if it describes the Earth, the Moon, or both. Write the word in the correct place on the diagram.

rotate	revolve	planet	satellite	orbit
air	water	crater	sphere	phases

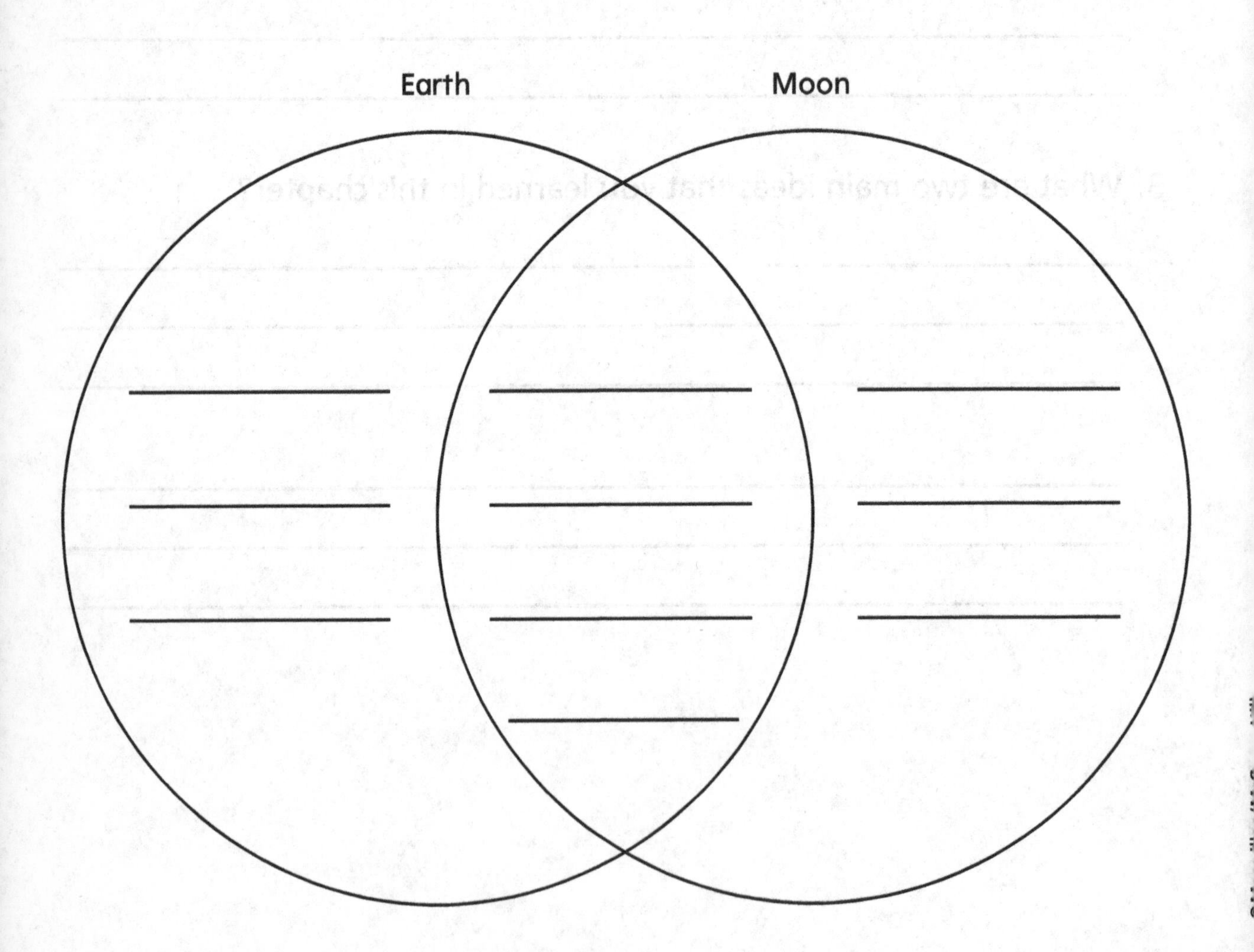

Name_______________________________ Date____________

Chapter Reading Skill

Chapter 8

Cause and Effect

A **cause** is something that happens, or an action. An **effect** is the result.

Read each cause below. Circle the effect.

1. Cause: The Earth makes a complete rotation in 24 hours. **Effect** **A.** There is day and night on Earth. **B.** We can see the Moon from Earth. **C.** The Earth has seasons.	**2. Cause:** The Earth, rotating on a tilted axis, revolves around the Sun every 365 days. **Effect** **A.** The Earth orbits the Sun. **B.** There is darkness on half of Earth. **C.** There are seasons on Earth.
3. Cause: Large chunks of rocks and metal fell on the Moon's surface. **Effect** **A.** The Moon has no air. **B.** The Moon has craters. **C.** The Moon has less gravity than Earth.	**4. Cause:** Telescopes make faraway objects appear larger and closer. **Effect** **A.** Scientists can study the planets. **B.** The planets are very far away. **C.** The Sun is the center of the solar system.
5. Cause: The Moon orbits the Earth. **Effect** **A.** The Moon is shaped like a sphere. **B.** The Earth orbits the Sun. **C.** We see phases of the Moon.	**6. Cause:** There is no air and water on the Moon. **Effect** **A.** People weigh less on the Moon. **B.** People cannot live on the Moon. **C.** The Moon is smaller than Earth.

Name ______________________ Date ____________

Chapter Reading Skill

Chapter 8

Weather Effects

Remember that a cause is something that happens. An effect is the result of that action.

Look at each picture below. Write the cause and effect for each picture.

Cause ______________________

____

Effect ______________________

____

Cause ______________________

____

Effect ______________________

____

Cause ______________________

____

Effect ______________________

____

Name________________________________ Date______________

Lesson Outline
Lesson 4

How Earth Moves

Fill in the blanks. Reading Skill: **Cause and Effect** - questions 5, 8, 13

What Causes Day and Night?

1. Earth's shape, which is like a ball, is called a(n) ____________________.
2. As Earth ____________________, there is daylight where Earth faces the Sun and darkness where Earth is turned away from the Sun.
3. It takes 24 hours for Earth to make one complete ____________________.
4. One complete rotation on Earth is one ____________________.
5. Earth rotates, or spins, around its ____________________.
6. At the north end of Earth's axis is the ____________________.
7. The tilt of Earth changes how the Sun's rays strike the ____________________.
8. During the day, the Sun rises very high in the sky at Earth's ____________________.

Name______________________ Date____________

Lesson Outline

Lesson 4

Fill in the blanks.

What Causes the Seasons?

10. An object that moves around another object ____________________.

11. Earth travels in a regular path around the Sun called a(n) ____________________.

12. It takes one year, or $365\frac{1}{4}$ days, for Earth to make one complete ____________________ around the Sun.

13. Many parts of the world have seasons because of Earth's tilted ____________________.

14. During the summer, North America is tilted ____________________ the Sun.

15. During the winter, North America is tilted ____________________ from the Sun.

How Does the Sun's Path in the Sky Change?

16. During the summer, in your part of the Earth, the Sun's path appears ____________________ in the sky.

17. During the winter, in your part of the Earth, the Sun's path appears ____________________ in the sky.

Name ______________________ Date __________

What Causes Day and Night?

You can learn a lot from diagrams and captions. This illustration shows why we have night and day on Earth.

The Rotating Earth

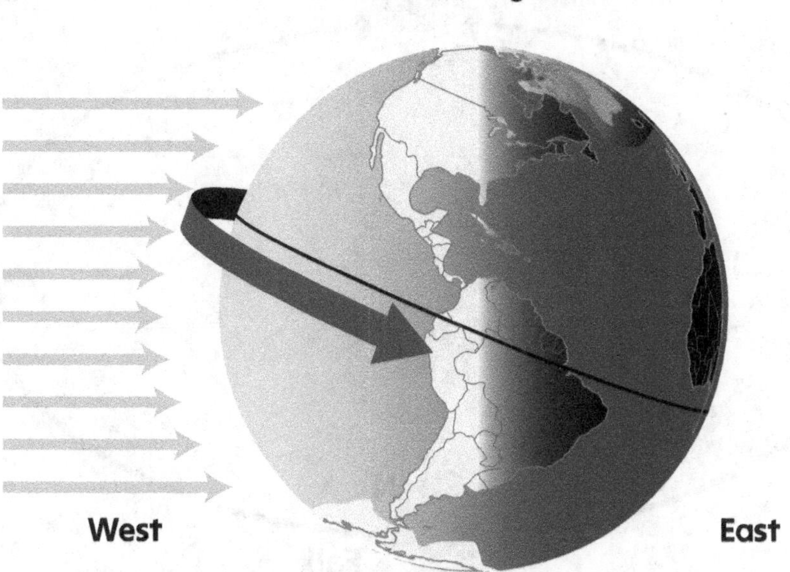

As Earth rotates, you see the Sun rise in the east and set in the west.

Answer these questions about the diagram above.

1. What is the title of this diagram? ______________________
2. In what direction does the Earth rotate? ______________________
3. What part of the Earth is in daytime in this picture?

____

Name______________________________ Date____________

Interpret Illustrations

Lesson 4

What Causes the Seasons?

This diagram shows why we have seasons on Earth.

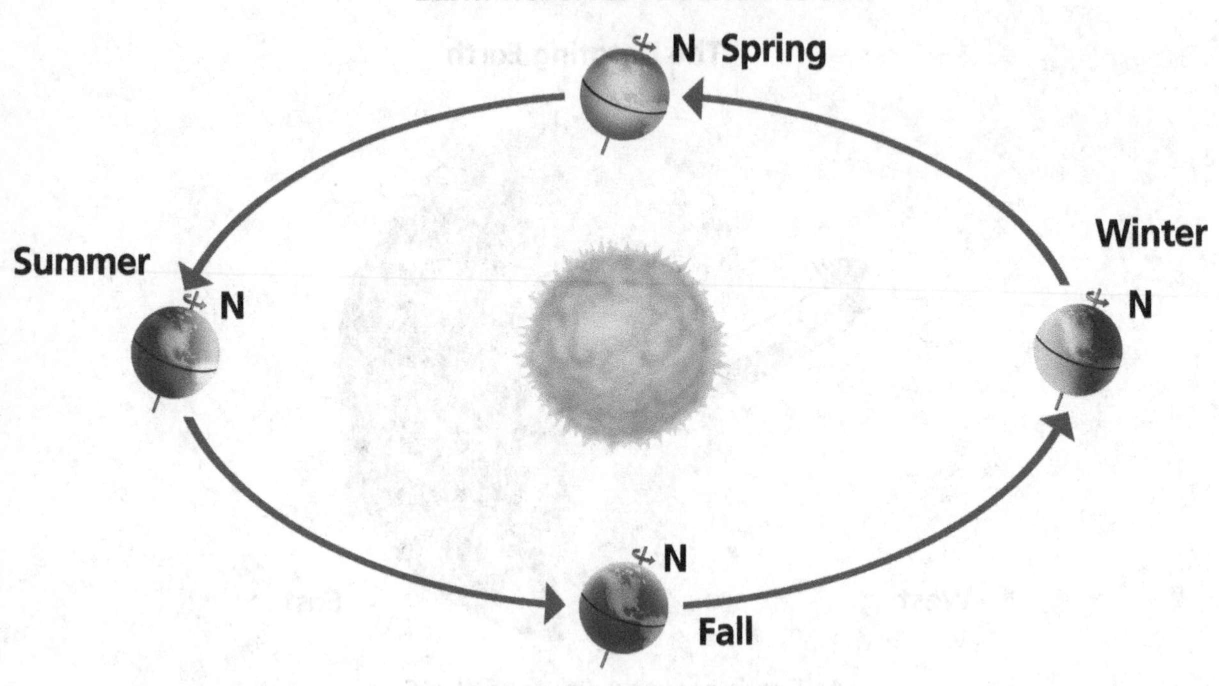

Answer these questions about the diagram above.

1. During which seasons is North America not tilted toward or away from the Sun? ______________________________

2. How is North America tilted during the summer season?
____

3. What happens to the temperature when North America is tilted away from the Sun? ______________________________

Use with textbook pages D38–D39

Name________________________________ Date_____________

Lesson Vocabulary
Lesson 4

How Earth Moves

Fill in the blanks.

1. Day and night happens on Earth because Earth ______________________, or spins around.
2. An imaginary line that goes through the center of Earth is its ______________________.
3. The Sun rises very high in the sky at Earth's ______________________ during the day.
4. During the winter, North America ______________________ away from the Sun.
5. Earth moves, or ______________________, around the Sun.
6. The path that an object follows as it travels around the Sun is called its ______________________.
7. Earth is shaped like a ball, or a ______________________.

Vocabulary
- sphere
- rotates
- axis
- revolves
- orbit
- equator
- tilts

Answer the question.

8. Why does the Sun seem to rise and set in the sky?

____

Name______________________________ Date______________

Cloze Test
Lesson 4

How Earth Moves

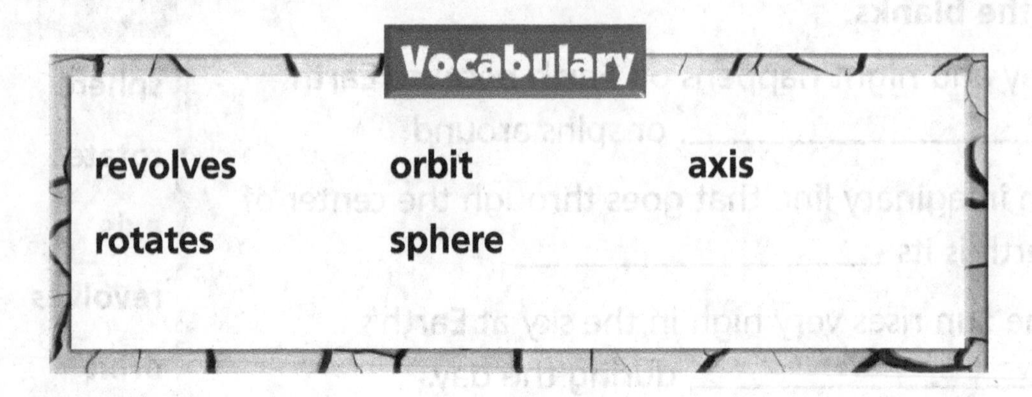

Fill in the blanks.

It is the movement of Earth that causes day and night. Earth is shaped like a ball, or a __________________. Earth __________________, causing daylight where Earth faces the Sun and darkness where Earth is turned away from the Sun. Earth rotates around an imaginary line drawn through its center called an __________________. Earth also __________________ around the Sun. It travels in an __________________, which is the path that an object follows as it revolves around another object.

Name________________________ Date____________

Lesson Outline
Lesson 5

Phases of the Moon

Fill in the blanks. Reading Skill: **Cause and Effect** - questions 3, 11

How Does the Moon's Shape Change?

1. The Moon is a(n) ____________________ of Earth.
2. Half of the Moon always faces toward the ____________________.
3. The Moon only appears to change shape because you see different amounts of its ______________________________ as it orbits Earth.
4. The Moon's changing shapes are called its ____________________.
5. The Moon passes through all of its phases in about ____________________ days.
6. The four phases of the Moon are
   a. ______________________________,
   b. ______________________________,
   c. ______________________________, and
   d. ______________________________.
7. No matter what the shape of the Moon may look like, the Moon is always a ____________________.

Name_______________________ Date___________

Lesson Outline

Lesson 5

Fill in the blanks.

How Are Earth and the Moon Different?

8. A day on the Moon lasts more than ______________ Earth days.
9. The Moon's gravity is about ______________ of Earth's gravity.
10. Most of the Moon's surface is covered with hollow areas, or pits, called ______________.
11. Some of the Moon's craters were formed by ancient ______________.
12. The Moon has no air or no liquid ______________.
13. Astronauts who visited the Moon wore spacesuits to protect them from the ______________.

How Does the Moon's Shape Change?

You can learn a lot from diagrams and captions. This diagram shows why the Moon appears to change its shape.

New Moon | First Quarter Moon | Full Moon | Last Quarter Moon

As the Moon orbits Earth, one side is always lighted. As the Moon changes position, different parts of its lighted side are seen from Earth. No matter what shape it may look like, the Moon is always a sphere.

Answer these questions about the diagrams above.

1. During which phase do you see the entire Moon lit up?

   ____

2. How does the shape of the Moon appear during the first quarter?

   ____

3. Describe what the Moon looks like during the New Moon phase.

   ____

4. What causes the different parts of the Moon to appear lighted?

   ____

   ____

Name ______________________ Date ____________

Interpret Illustrations

Lesson 5

How Are Earth and the Moon Different?

Think about how the Earth and the Moon are alike and how they are different.

Large and small craters cover the Moon's rocky surface.

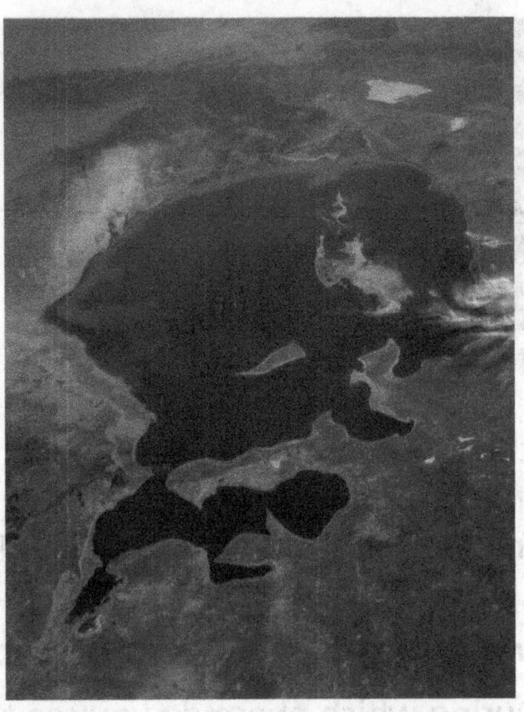

Land and water cover the Earth's surface.

Answer these questions about the diagram above.

1. What does the first picture show? ______________________
2. How would you describe the Moon's surface? ______________________

   ____
3. How would you describe the Earth's surface? ______________________

   ____
4. Why do you think that life cannot exist on the Moon? ______________________

   ____

Name______________________________ Date______________

Lesson Vocabulary

Lesson 5

Phases of the Moon

Fill in the blanks.

Vocabulary
satellite
phases
New Moon
Last Quarter
revolve
rotate
craters

1. It takes about 29 days for the Moon to ______________________ around Earth.
2. Hollow areas, or pits on the Moon's surface are called ______________________.
3. You only see about half the Moon during it's ______________________ phase.
4. An object that orbits another larger object in space is called a(n) ______________________.
5. As the Moon orbits Earth, it appears to have changing shapes called ______________________.
6. It takes the Moon more than 27 Earth days to ______________________ all the way around.
7. You cannot see the Moon during the ______________________ phase.

Answer the question.

8. How are craters formed on the Moon?

____

____

Name________________________________ Date______________

Cloze Test
Lesson 5

Phases of the Moon

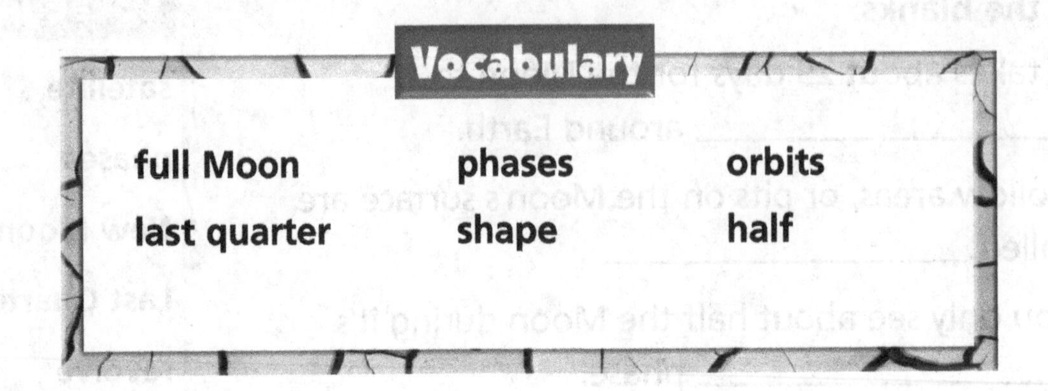

Fill in the blanks.

The Moon is called Earth's satellite because it ____________________ Earth. The Moon only appears to change its ____________________. The changing shapes of the Moon are called its ____________________. During the ____________________, the entire Moon appears to be lit up. During the ____________________, only half of the Moon appears to be lit up. Like Earth, ____________________ of the Moon always faces the Sun.

Name______________________ Date____________

Lesson Outline
Lesson 6

The Sun and the Planets

Fill in the blanks. Reading Skill: **Cause and Effect** - questions 6, 7, 9

What Is the Solar System?

1. The Sun and all the objects that orbit it is called the ________________________.
2. There are nine ________________ that orbit the Sun.
3. The nine planets are, in order from the Sun,
   a. ________________,
   b. ________________,
   c. ________________,
   d. ________________,
   e. ________________,
   f. ________________,
   g. ________________,
   h. ________________, and
   i. ________________.
4. The Sun is a ________________.
5. The Sun is so big that if it were hollow, more than one ________________ Earths would fit inside it.
6. The Sun is so far away that it takes ________________ eight minutes to get to Earth.
7. Earth gets its heat and ________________ from the Sun.

Name ______________________ Date ______________

Lesson Outline
Lesson 6

Fill in the blanks.

What Are the Nine Planets Like?

8. All of the nine planets ______________ sunlight from the Sun.
9. The name planet comes from a word that means ______________.
10. Each planet moves in a different orbit and at a different ______________.
11. The planets are divided into two groups called the inner planets and the ______________.
12. All of the inner planets are small and made up of solid, ______________ materials.
13. All of the outer planets are made up mostly of ______________, except for Pluto.

How Do We Learn About Space?

14. A tool that gathers light with mirrors and lenses is called a ______________.
15. A curved piece of glass is called a(n) ______________.

Name________________________________ Date______________

Interpret Illustrations

Lesson 6

The Inner Planets

You can learn a lot from illustrations and captions. These illustrations show the inner planets of the solar system.

Mercury is the closest planet to the Sun. It looks a lot like Earth's Moon.

Space probes have traveled to cloud-covered Venus many times. Photographs of its surface show features such as mountains and canyons.

Earth is our home. It is the only planet with liquid water.

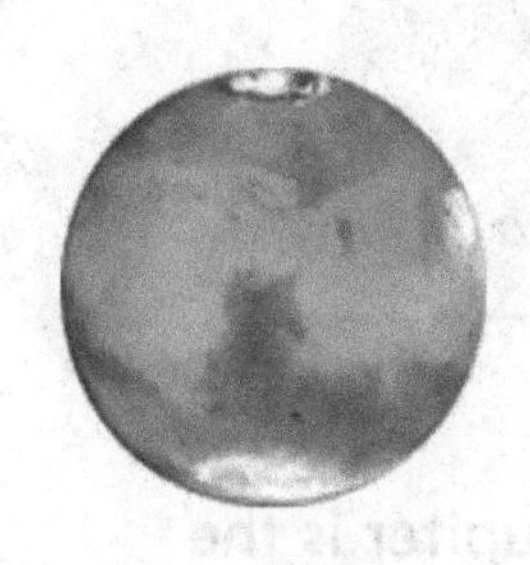

Mars has some water, but most of it is frozen ice. Mars is known as the "red planet" because of its reddish rocks and soil.

Answer these questions about the illustrations above.

1. Which planet is the only planet with liquid water? ______________
2. Which cloud-covered planet has been visited by space probes?

   ____
3. Which planet is known as the "red planet?" ______________
4. What is Mercury's position in the solar system?

   ____

Name ____________________ Date ____________

Interpret Illustrations

Lesson 6

The Outer Planets

These illustrations show the outer planets of the solar system.

Jupiter is the largest planet in our solar system. The Great Red Spot on Jupiter has been whirling around for 300 years.

Saturn is known for its thousands of beautiful rings. They are made up of different-sized bits of ice and rock that orbit the planet.

Uranus is called the "sideways planet" because it rotates on its side.

Neptune is more than two billion miles from Earth. It has a Great Dark Spot similar to Jupiter's Great Red Spot.

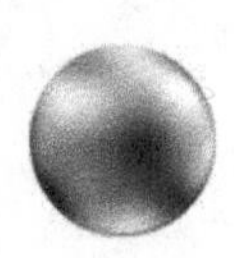

Little is known about Pluto because it is so far away.

Answer these questions about the illustrations above.

1. Which of the outer planets rotates on its side? ____________________
2. Which planet is the largest in our solar system? ____________________
3. What are Saturn's rings made of? ____________________
4. Which two planets have great big spots on them? ____________________

Use with textbook page D57

Name____________________ Date__________

Lesson Vocabulary
Lesson 6

The Sun and Its Planets

Fill in the blanks.

Vocabulary
solar system
planet
star
telescope
lens
inner planets
outer planets

1. A tool that makes faraway objects appear larger, closer, and clearer is called a(n) ____________________.
2. The first four planets that are the closest to the Sun are called the ____________________.
3. A hot, glowing ball of gases is called a(n) ____________________.
4. The Sun and all the objects that orbit it make up the ____________________.
5. The last five planets that are farthest from the Sun are called the ____________________.
6. Telescopes gather light with mirrors and a curved piece of glass called a(n) ____________________.
7. A large body of rock or gas that orbits the Sun is called a(n) ____________________.

Answer the question.

8. How do the inner planets differ from the outer planets?

____

____

Name____________________________ Date______________

Cloze Test
Lesson 6

The Sun and Its Planets

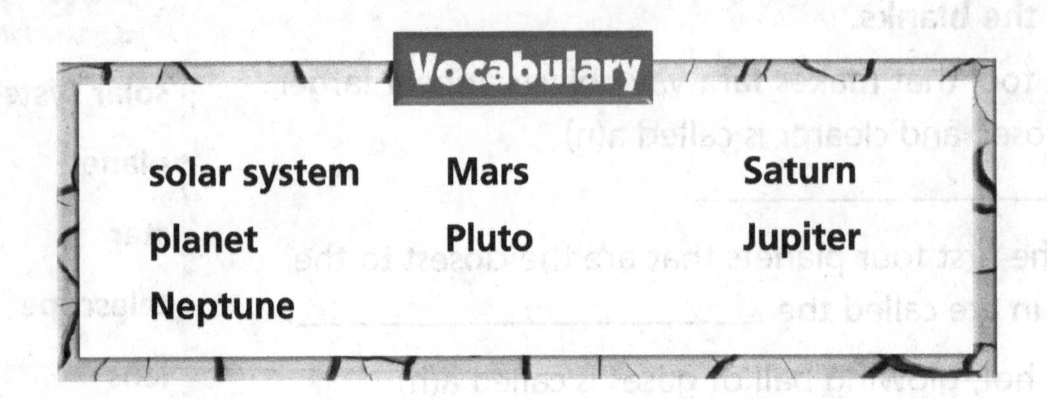

Fill in the blanks.

The ______________________ is made up of the Sun and all the objects that orbit it. A ______________________ is a large body of rock or gas that orbits the Sun. The planet that is known for its beautiful rings is _________________. The small planet at the very end of the solar system that scientists do not know much about is ______________________. The Great Red Spot is found on the planet ______________________. The Great Dark Spot is found on ______________________. A planet that has some frozen water on it is ______________________.

Name__________________________ Date______________

Chapter Vocabulary

Chapter 8

Earth in Space

Circle the letter of the best answer.

1. Earth ______________________, or spins around, on its axis.
   - a. revolves.
   - b. orbits.
   - c. rotates.
   - d. tilts.

2. Earth makes one complete rotation in
   - a. 24 hours.
   - b. 365 days.
   - c. 15 days.
   - d. 9 hours.

3. Earth travels in a regular path around the Sun called a(n)
   - a. axis.
   - b. orbit.
   - c. rotation.
   - d. pole.

4. An imaginary line drawn through the center of Earth around which Earth rotates is called its
   - a. axis.
   - b. equator.
   - c. poles.
   - d. none of the above.

5. An object that orbits another larger object in space is called a(n)
   - a. telescope.
   - b. satellite.
   - c. Moon.
   - d. phase.

6. Craters are formed by all of these EXCEPT
   - a. chunks of rock.
   - b. volcanoes.
   - c. chunks of metal.
   - d. planets.

7. All of the following are phases of the Moon EXCEPT
   a. new Moon. b. old Moon.
   c. full Moon. d. first quarter.

8. You would weigh less on the Moon than on Earth because the Moon has less
   a. gravity. b. air.
   c. mass. d. water.

9. A large body of rock or gas that orbits the Sun is called a(n)
   a. star. b. solar system.
   c. planet. d. all of the above.

10. One of the inner planets is
   a. Neptune. b. Pluto.
   c. Jupiter. d. Mars.

11. In the order of planets from the Sun, Earth's position is
   a. first. b. third.
   c. second. d. fourth.

12. A tool that scientists use to view the planets is called a
   a. telescope. b. microscope.
   c. lens. d. satellite.

Name________________________________ Date____________

Unit Vocabulary
Unit D

Vocabulary Match

Write the letter of the correct meaning next to each vocabulary word.

_____ **1.** axis	**a.** a shape like a ball
_____ **2.** sphere	**b.** to turn
_____ **3.** phases	**c.** an imaginary line through Earth's center
_____ **4.** solar system	**d.** to move around another object
_____ **5.** revolve	**e.** a path an object follows as it revolves
_____ **6.** planet	**f.** an object that orbits another larger object in space
_____ **7.** satellite	**g.** the different shapes the Moon appears to have
_____ **8.** orbit	**h.** a hollow area, or pit in the ground
_____ **9.** star	**i.** a large body of rock or gas that orbits the Sun
_____ **10.** telescope	**j.** the Sun and all the objects that orbit it
_____ **11.** rotate	**k.** a hot, glowing ball of gas
_____ **12.** lens	**l.** a tool that makes faraway objects appear larger and closer
_____ **13.** crater	**m.** a curved piece of glass

14. Choose two words from the list above. Use each word in a sentence.

____

____

____

____

Name ______________________________ Date ____________

Unit Vocabulary

Unit D

Planet Wheel

Start with the letter E. Write that letter on the first line below the picture of Earth. Next write every third letter on the rest of the lines. (The next letter has the Moon next to it to help you.) Continue around the wheel. After you go around three times, you will find out a fact about Earth.

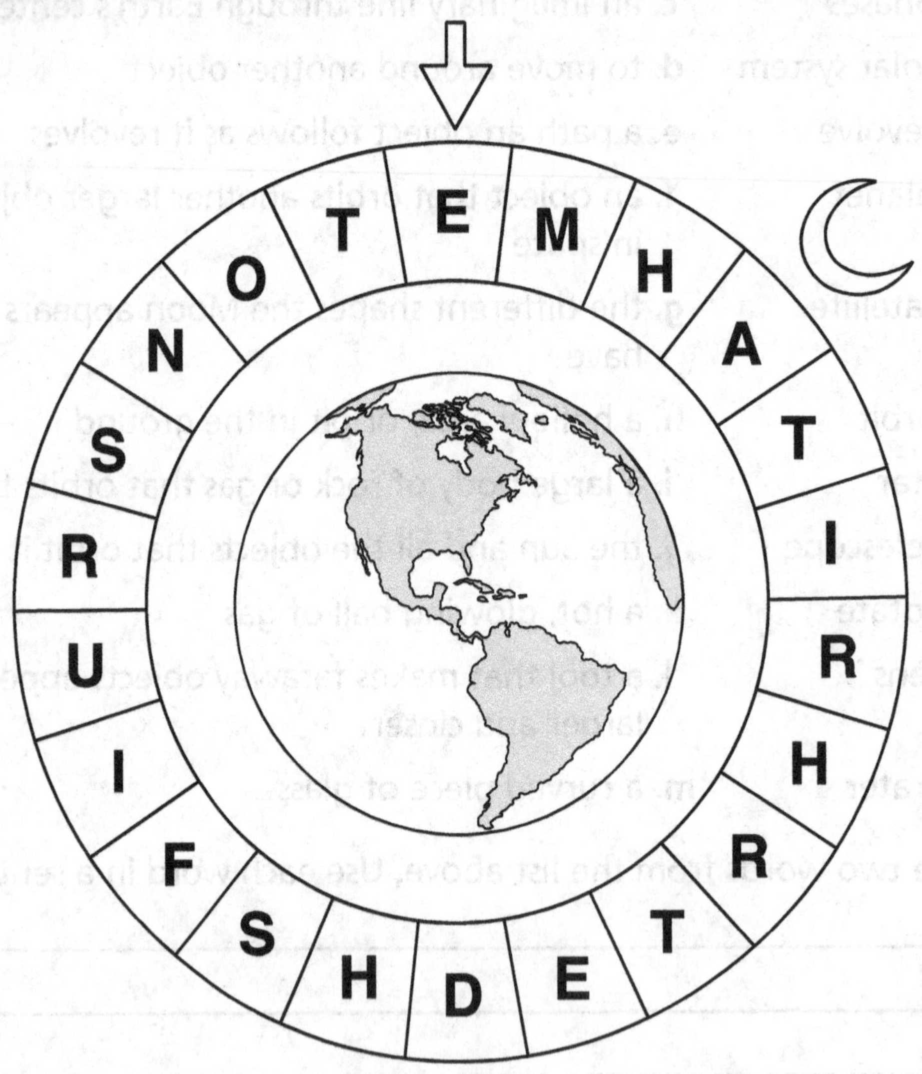

_ _

Name______________________________ Date______________

Planet Crossword

Read each clue. Write the answer in the crossword puzzle.

Across

1. the center of our solar system
4. the planet with the Great Red Spot
5. a satellite of Earth
6. the planet with beautiful rings
7. the second planet from the Sun
9. the planet that rotates on its side
10. the only planet with liquid water

Down

2. the planet with the Great Dark spot
3. the "Red Planet"
5. the closest planet to the Sun
8. an outer planet that is made of rock and frozen gases

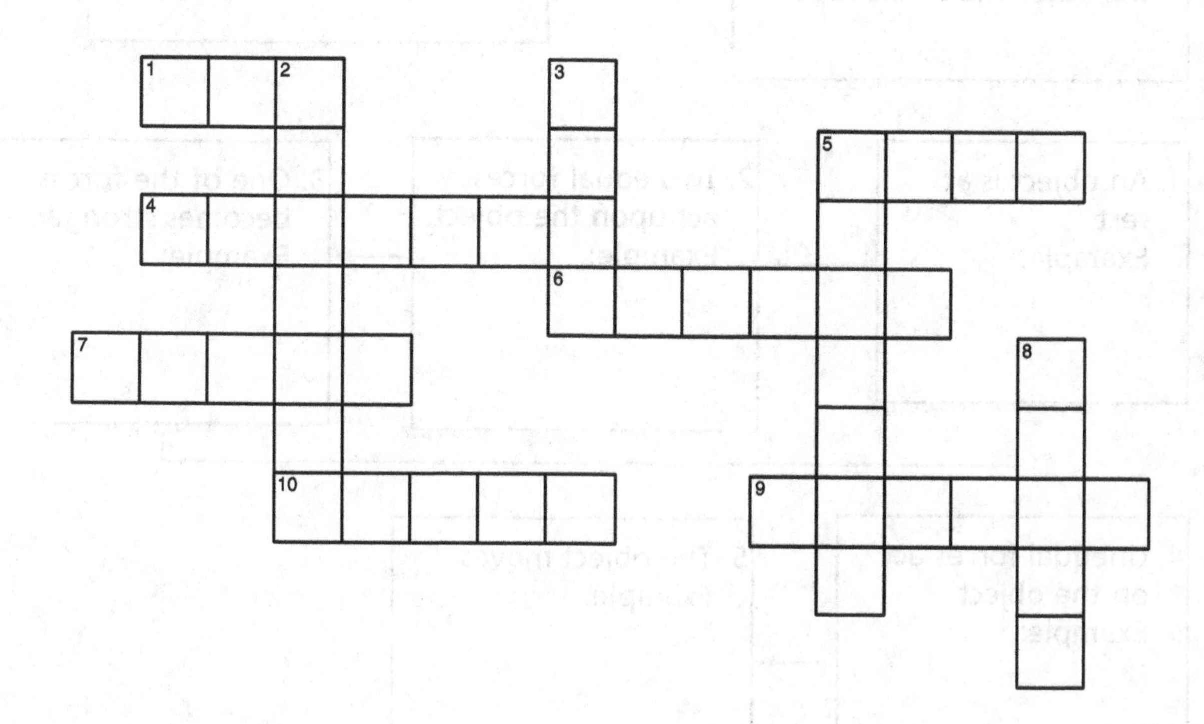

Name ______________________ Date ____________

How Things Move

Change in motion can occur in several steps. Below is a diagram showing the steps of a change in motion. An example of each step is provided. **Look at the top diagram carefully, and then give similar examples of the same steps in the bottom diagram.**

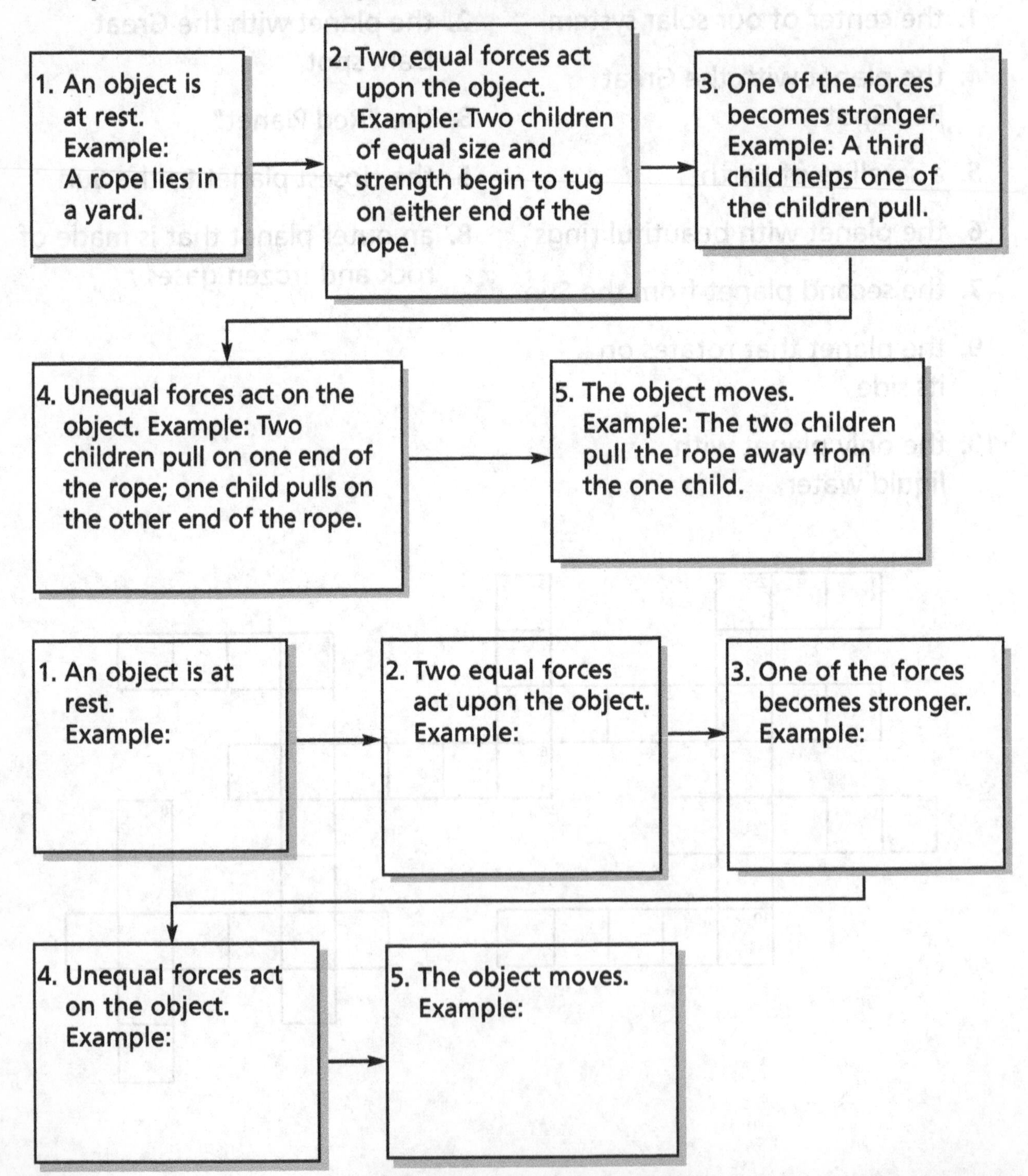

Name________________________ Date____________

Chapter Reading Skill
Chapter 9

Main Idea and Supporting Details

A topic sentence tells the **main idea** of a paragraph. **Read each paragraph. Write the letter of the topic sentence that makes the most sense.**

1. _____ Once you start going down a hill, you will go faster and faster. You can stop if you know how to use your skates. Or, you just might get stopped by a tree!

   A. In-line skating is a popular sport.

   B. It is important to know how to stop when you are in-line skating.

   C. In-line skaters need to wear pads.

2. _____ Two teams line up on opposite sides of a rope. The rope might hang over a puddle of water. The teams pull on the rope. The losing team gets pulled into the water.

   A. A tug-of-war can be a lot of fun.

   B. Ropes are made of very strong materials.

   C. This is how you play tug-of-war.

3. _____ In a space shuttle, there is no pull of gravity. Food and water can float in the air. Astronauts strap themselves into sleeping bags when they sleep. This way, they don't float all over the space shuttle.

   A. Astronauts need to get used to floating in space.

   B. Astronauts orbit Earth.

   C. Many people would like to travel in space.

Name ______________________ Date ____________

Chapter Reading Skill

Chapter 9

A Step in Time

Read the story. Answer the questions.

June 20, 1969, was a very important day in space exploration. On that day, astronauts Neil Armstrong and Buzz Aldrin landed their spacecraft on the Moon. Neil Armstrong opened the spacecraft door and climbed down the ladder. He stepped onto the Moon. This was the first time a human being ever set foot on the Moon.

He bounced around the Moon's surface easily. He could do this because he felt so light. On the Moon, he weighed much less than on Earth. That is because the pull of gravity is greater on Earth than on the Moon.

Later, Buzz Aldrin joined Neil. The two astronauts took many pictures. They also collected rock samples from the Moon. Scientists on Earth could study these later to find out more about the Moon. Before they left, the two astronauts put the flag of the United States on the Moon. This was a day that people would never forget.

1. Which sentence tells the main idea of the story?
   _____ Neil Armstrong bounced around the Moon.
   _____ June 20, 1969, was an important day in space exploration.
   _____ Neil Armstrong and Buzz Aldrin were astronauts.
   _____ The pull of gravity is greater on Earth than on the Moon.
2. List three details that support the main idea.

____

____

____

____

Name ______________________________ Date ______________

Lesson Outline

Lesson 1

Motion and Speed

Fill in the blanks. Reading Skill: **Main Idea and Supporting Details** - questions 5, 6, 7

How Do You Know If Something Has Moved?

1. Position is the ______________ of an object.
2. To describe an object's position, you ______________ it to the positions of other objects.
3. ______________ is the length between two places.
4. Measuring the length between the ______________ and ______________ positions of an object gives you distance.

How Do You Measure Motion?

5. Since the cheetah in the diagram on textbook pages E8 and E9 changes position, it is in ______________.
6. Motion can include a change in ______________.
7. In the diagram, the motion stops when the cheetah no longer changes ______________.
8. ______________ is how fast an object moves over a certain distance.
9. To measure speed, you need to measure ______________ and ______________.
10. A fast-moving object goes a long distance in a(n) ______________ period of time.
11. If a cheetah could run for one hour at its fastest speed, it would travel ______________.
12. Speed is the distance ______________ by the time.

Name ______________________ Date ____________

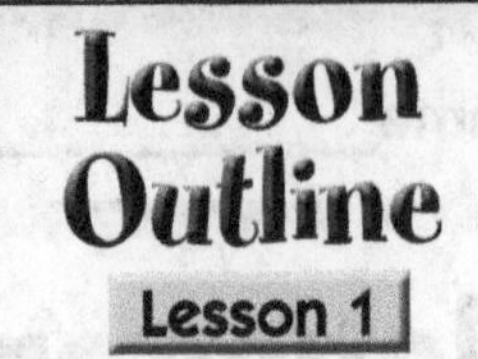

What Do Maps Tell You?

13. A map is a flat ______________ that shows where objects are located.

14. Maps show the ______________ of things.

15. You use ______________ to read a map.

16. Directions include north, ______________, east, and ______________.

17. You use a(n) ______________ to show what different symbols stand for.

Name________________________________ Date______________

Interpret Illustrations

Lesson 1

How Do You Measure Motion?

A diagram uses pictures and words to describe a thing or process. This diagram shows a cheetah in motion. Each numbered part of the diagram illustrates a fact about motion.

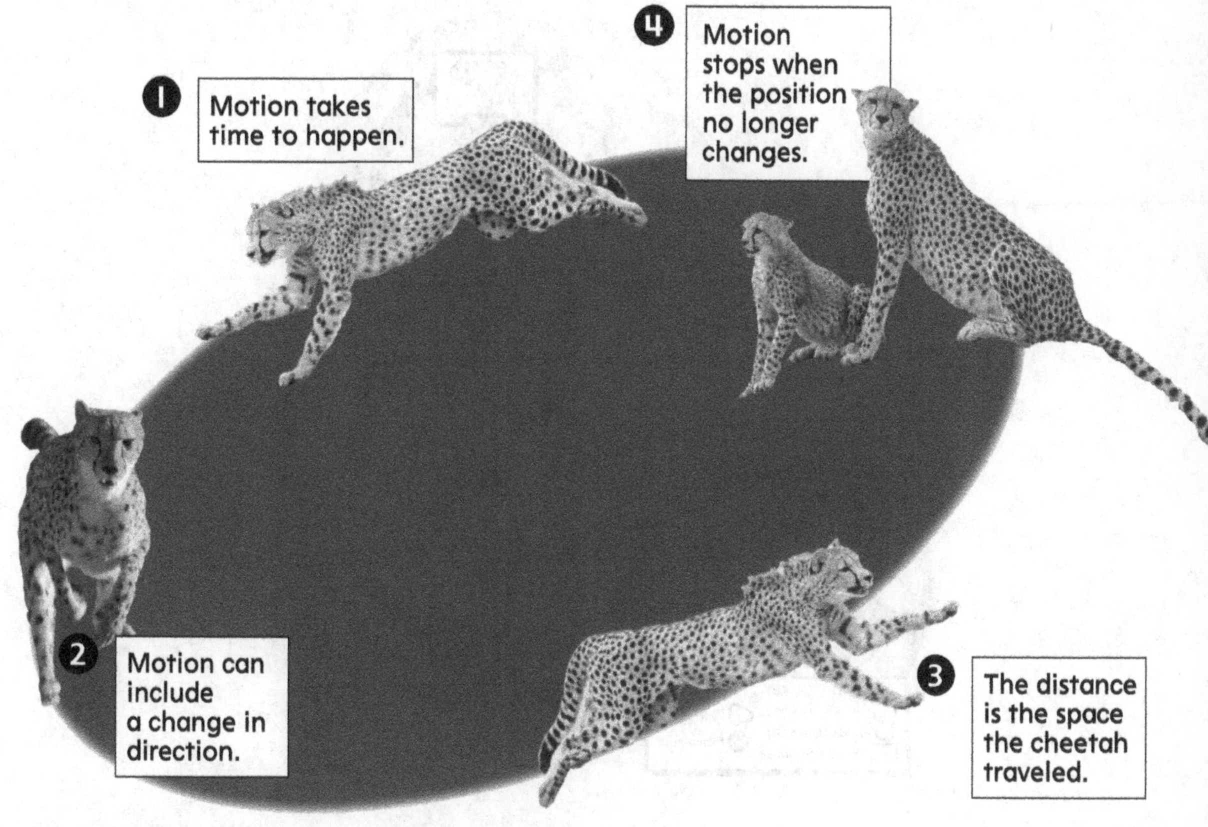

Use the diagram to answer the questions.

1. What fact about motion is given in Sentence 1?

____

2. According to Sentence 3, what is distance?

____

3. According to Sentence 4, what happens when the position of a moving object no longer changes?

____

4. According to Sentence 2, what can motion include?

____

Name ______________________ Date __________

Interpret Illustrations

Lesson 1

What Do Maps Tell You?

A map shows if one place is north, south, east, or west of another place. This map shows the areas in a zoo.

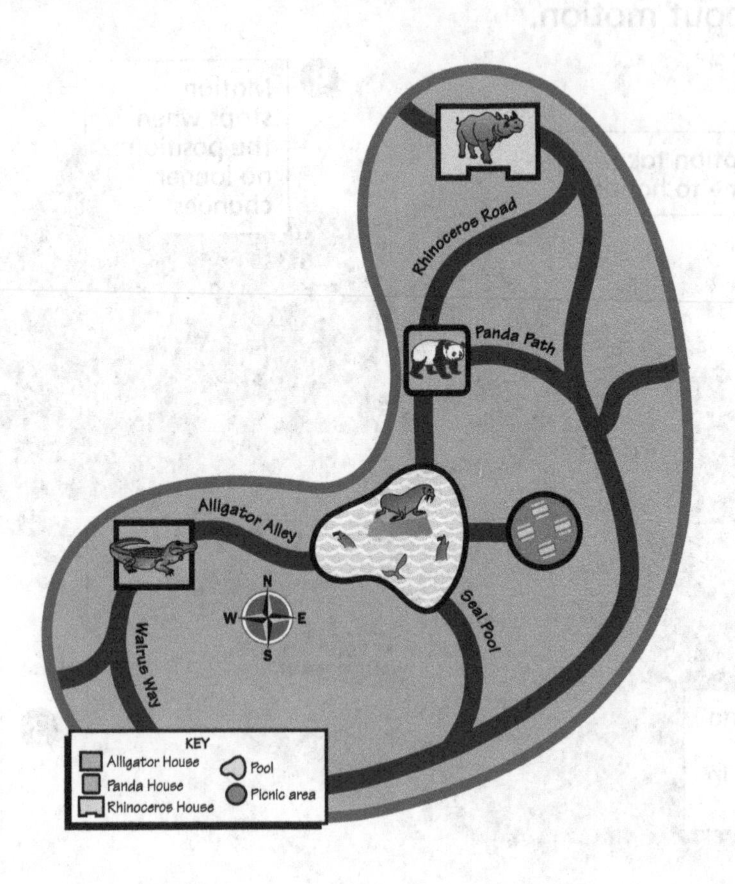

Fill in each blank with one of these words: north, south, east, or west.

1. The Panda House is ________________ of the pool.
2. The Alligator House is ________________ of the pool.
3. The picnic area is ________________ of the Alligator House.
4. The pool is ________________ of the Rhinoceros House.
5. The Panda House is ________________ of the Rhinoceros House.
6. The picnic area is ________________ of the pool.
7. The Rhinoceros House is ________________ of the Panda House.

Name____________________ Date__________

Lesson Vocabulary
Lesson 1

Motion and Speed

Vocabulary
key
motion
stops
position
time
distance
objects
speed
direction
map

Fill in the blanks.

1. The location of an object is its ________________.
2. You describe an object's position by comparing it with the positions of other ________________.
3. An object is in ________________ if it changes position.
4. Motion happens over a period of ________________.
5. Motion can include a change of ________________.
6. When the position of an object no longer changes, motion ________________.
7. ________________ is how fast an object moves over a certain distance.
8. To measure speed, you need to measure time and ________________.
9. A flat drawing that shows where objects are located is a(n) ________________.
10. A map ________________ tells you what the different symbols stand for.

Name ______________________ Date ____________

Cloze Test
Lesson 1

Motion and Speed

Vocabulary

moves	**location**	**faster**
distance	**postion**	

Fill in the blanks.

When an object changes ______________, it is in motion. Position is the ______________ of an object. You can measure speed to see how fast an object ______________. Your running speed is ______________ than your walking speed. To measure your speed, you need to know the ______________ you travel in a period of time.

Name______________________________ Date______________

Lesson Outline
Lesson 2

Forces

Fill in the blanks.  Reading Skill: **Main Idea and Supporting Details** - questions 5, 6

What Are Pushes and Pulls?

1. All pulls and pushes are ________________.
2. The motion of an object can often be ________________ by a force.
3. Forces always work in ________________.
4. A force can make an object start moving, ________________ moving, or change direction.
5. You have to push or pull a heavier object ________________ to make it move.
6. If you push on something, you can feel the object ________________ back.

What Force Is Always Pulling on You?

7. The force of ________________ keeps objects pulled toward the Earth.
8. Things fall to Earth because they are ________________ by Earth's gravity.

Name______________________________ Date______________

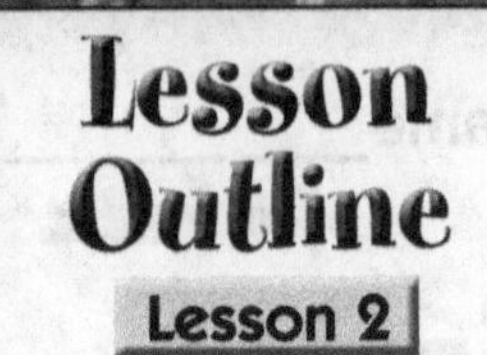

What Is Weight?

9. Weight is how much pull ________________ has on an object.
10. You can measure how heavy or light an object is by measuring its ________________.
11. The weight of an object is about the same anywhere on Earth because the pull of ________________ is about the same.
12. Since the pull of gravity is different on each ________________, objects have different weights away from Earth.
13. Both newtons and pounds are units of ________________ used to measure weight.

Name ____________________ Date ______________

Interpret Illustrations

Lesson 2

What Are Pushes and Pulls?

Pictures and words together can describe a thing or process. These pictures and words explain facts about forces. Read each numbered box.

Box 1
All pushes and pulls are forces. Pushes move away from you. Pulls move toward you.

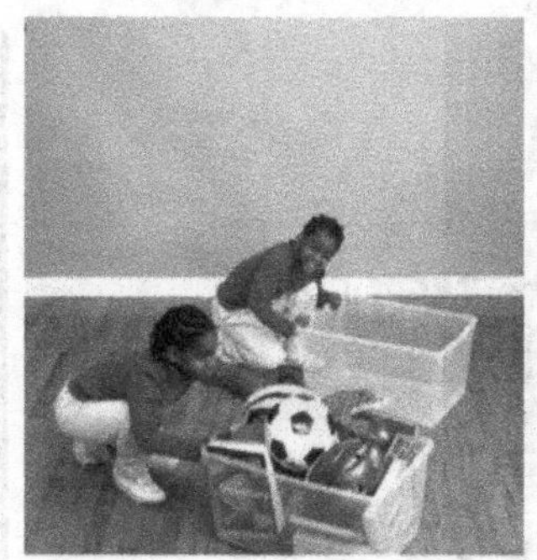

Box 2
Forces may change the motion of an object. The heavier an object, the more force you need to move it.

Box 3
When you push or pull on something, it pushes or pulls on you. This force that you feel is a force in the opposite direction.

Box 4
Many things can create forces. Magnets can push or pull on objects without even touching them.

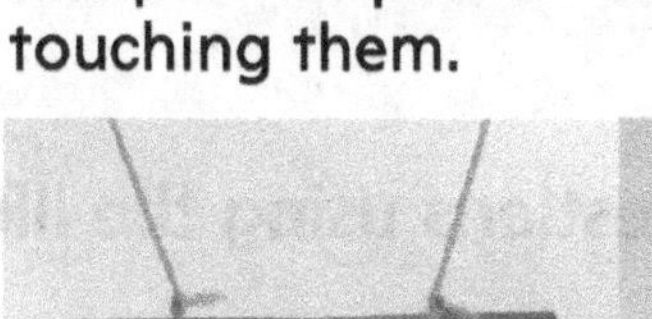

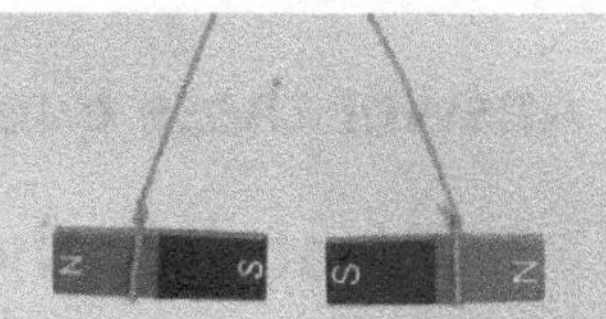

Use the pictures and words above to answer these questions.

1. Box 1 compares pushes and pulls. How are they alike and different?

____

2. Look at Box 2. What clue from the picture shows which box is heavier?

____

____

3. What force is acting on the girl in Box 3?

____

4. Box 4 describes a certain kind of force. How is this force described?

____

Name______________________________ Date_______________

Interpret Illustrations

Lesson 2

What Is Weight?

Illustrations often give us information.

Answer these questions using the illustrations above.

1. What does the first picture show?

____

2. How much do the apples weigh in pounds?

____

3. How much do the apples weigh in newtons?

____

4. Is the weight of the apples the same in pounds as it is in newtons? Explain.

____

5. What does the second picture show?

____

6. What does this picture tell you about the pull of gravity on the Moon? Is it the same or different from the pull of gravity on Earth?

____

Name______________________________ Date____________

Forces

Vocabulary

- measure
- pairs
- forces
- toward
- gravity
- force
- away from
- motion
- weight

Fill in the blanks.

1. All pushes and pulls are ____________________.
2. A force can often change the ____________________ of an object.
3. A newton is a unit used to ____________________ weight.
4. A push is a force that moves ____________________ you.
5. A pull is a force that moves ____________________ you.
6. Forces always work in ____________________.
7. The heavier an object is, the more ____________________ needed to move it.
8. The pulling force between two objects is called ____________________.
9. The pull of gravity on an object is the object's ____________________.

Name________________________________ Date______________

Cloze Test
Lesson 2

Forces

Vocabulary

gravity	**weight**	**newtons**
planet	**pounds**	

Fill in the blanks.

You can measure the ________________ of objects to find out how heavy or light they are. The pull of ________________ on an object is its weight. This pull is different on each ______________. Scientists use the metric system to measure forces such as weight in units called ________________. In the English system of measurement, ________________ are the units used to measure the same forces.

Name______________________________ Date____________

Lesson Outline
Lesson 3

Changes in Motion

Fill in the blanks.  Reading Skill: **Main Idea and Supporting Details** - questions 1, 2, 3, 4, 5, 8

What Causes a Change in Motion?

1. The diagram on textbook page E24 shows how unbalanced ________________ create a change of motion in a tug-of-war.
2. In the top part of the diagram, nothing moves because both sides ________________ equally.
3. The bottom part of the diagram shows that unequal forces cause a change in the ________________ of the rope.
4. A change in an object's motion is the result of all the ________________ that are acting on the object.
5. The diagram on textbook page E25 shows different types of ________________ in motion.
6. The diagram shows that a body at rest can start ________________.
7. A moving body can speed up, change ________________, or slow down.
8. The last part of the diagram shows that a body can ________________ moving.

Name ______________________ Date ____________

Lesson Outline

Lesson 3

Fill in the blanks.

Why Do Things Stop Moving?

9. Friction is a force that occurs when one object ______________ against another object.
10. A ball rubbing across a floor creates ______________.
11. A great deal of friction is produced by ______________ materials.
12. Even though rubber is ______________, it produces a lot of friction.
13. When a brake pad presses against the rim of a bicycle wheel, friction ______________ down the wheel and the bicycle.

How Can You Control Friction?

14. While you can't get rid of friction, you can change the ______________ of friction you have.
15. Slippery things can be used to ______________ friction.
16. Rough or sticky things can be used to ______________ friction.

Name________________________________ Date______________

Interpret Illustrations

Lesson 3

What Causes a Change In Motion?

A diagram uses pictures and words to describe a thing or process. This diagram shows a skier in motion. Each numbered part of the diagram illustrates a change in a moving body.

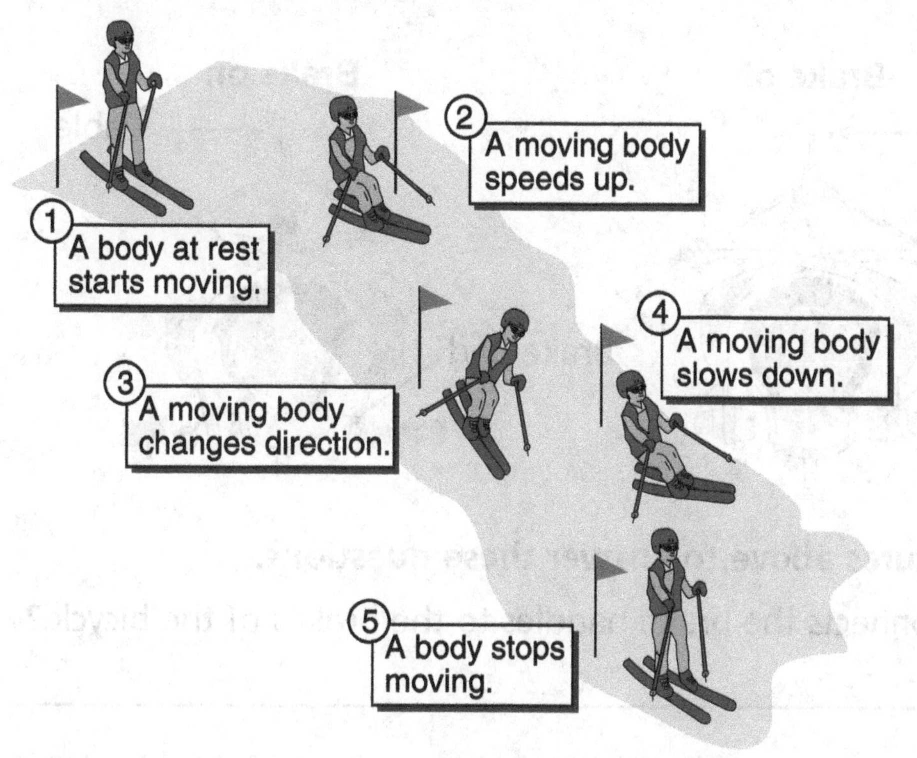

Use the diagram above to answer these questions.

1. Which part of the diagram shows an object at rest?

____

2. Which part of the diagram shows an object changing direction?

____

3. Which part of the diagram shows an object slowing down?

____

4. Which part of the diagram shows an object speeding up?

____

5. Which part(s) of the diagram shows an object in motion?

____

6. Which part(s) of the diagram shows a change in an object's motion?

____

Name______________________________ Date____________

Interpret Illustrations

Lesson 3

Why Do Things Stop Moving?

Diagrams often tell us how something works. This diagram shows the brakes, cables, and brake handles of a bicycle. It also shows how the brake pads look when they are on and off.

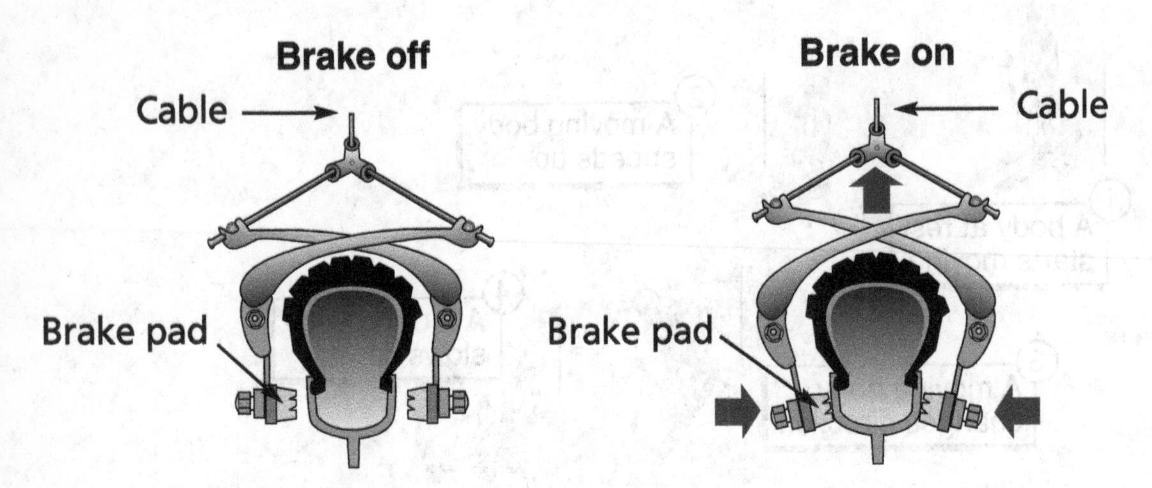

Use the pictures above to answer these questions.

1. What connects the brake handles to the brakes of the bicycle?

____

2. How many brake pads are on the front wheel of this bicycle?

____

3. What do the brake pads look like when the brake is off?

____

____

4. What happens to the brake pads when the brakes of the bicycle are on?

____

Name______________________ Date____________

Lesson Vocabulary
Lesson 3

Changes in Motion

Fill in the blanks.

Vocabulary
motion
more
increase
forces
direction
reduce
friction
speed
balanced
unbalanced

1. When an object that is resting starts to move, there is a change in ______________.
2. A change in motion occurs when a moving object speeds up, slows down, changes ______________, or stops.
3. A change in an object's motion is the result of all the ______________ that are acting on the object.
4. When the forces acting on an object are ______________, the object does not move.
5. When the forces acting on an object are ______________, the object moves.
6. The force that occurs when one object rubs against another object is ______________.
7. Friction slows down an object's ______________.
8. Rough materials produce ______________ friction than smooth materials.
9. People use slippery things to ______________ friction.
10. People use rough or sticky things to ______________ friction.

Name________________________ Date____________

Cloze Test
Lesson 3

Changes in Motion

Vocabulary

pad	**increase**	**reduce**
stops	**friction**	

Fill in the blanks.

________________ is a force that occurs when an object rubs against another object. To ________________ friction, people use slippery things. They use rough or sticky things to ________________ friction. When you squeeze the brake lever on your bike, there is friction between the brake ________________ and the rim of the wheel. The wheel slows down and then the bike ________________.

Name________________________ Date____________

Chapter Vocabulary

Chapter 9

How Things Move

Circle the letter of the best answer.

1. The location of an object is its
   a. distance.
   b. position.
   c. speed.
   d. weight.

2. When an object changes position, it is in
   a. gravity.
   b. location.
   c. motion.
   d. speed.

3. How fast an object moves is its
   a. direction.
   b. location.
   c. motion.
   d. speed.

4. The motion of an object can be changed with a
   a. force.
   b. map.
   c. motion.
   d. speed.

5. All forces work
   a. against you.
   b. alone.
   c. in groups.
   d. in pairs.

6. Things fall to Earth because they are pulled by Earth's
   a. gravity.
   b. speed.
   c. motion.
   d. weight.

Circle the letter of the best answer.

7. To reduce friction, people use
   a. rough or sticky things. b. heavy things.
   c. slippery things. d. rubber pads.

8. You can find how heavy or light something is by measuring its
   a. distance. b. height.
   c. speed. d. weight.

9. Scientists measure weight in
   a. inches. b. newtons.
   c. gravity. d. centimeters.

10. Since the pull of gravity on the Moon is different from that on Earth, you would
    a. weigh nothing on the Moon. b. not weigh the same on both.
    c. weigh more on the Moon. d. weigh the same on both.

11. An object changes its motion when the forces acting on it are
    a. balanced. b. equal.
    c. removed. d. unbalanced.

12. A force that occurs when one object rubs against another object is
    a. pushes. b. gravity.
    c. friction. d. pulls.

Name________________________________ Date______________

Chapter Summary

Chapter Summary

1. What is the name of the chapter you just finished reading?

____

____

2. What are two vocabulary words you learned in the chapter? Write a definition for each.

____

____

____

3. What are two main ideas that you learned in this chapter?

____

____

____

____

____

____

Name_______________________________ Date_______________

Chapter Graphic Organizer
Chapter 10

Work and Machines

Many tools you use are made of simple machines. Fill in the boxes in the concept map to organize information about simple machines. You may use the Chapter in your book, Work and Machines, for help.

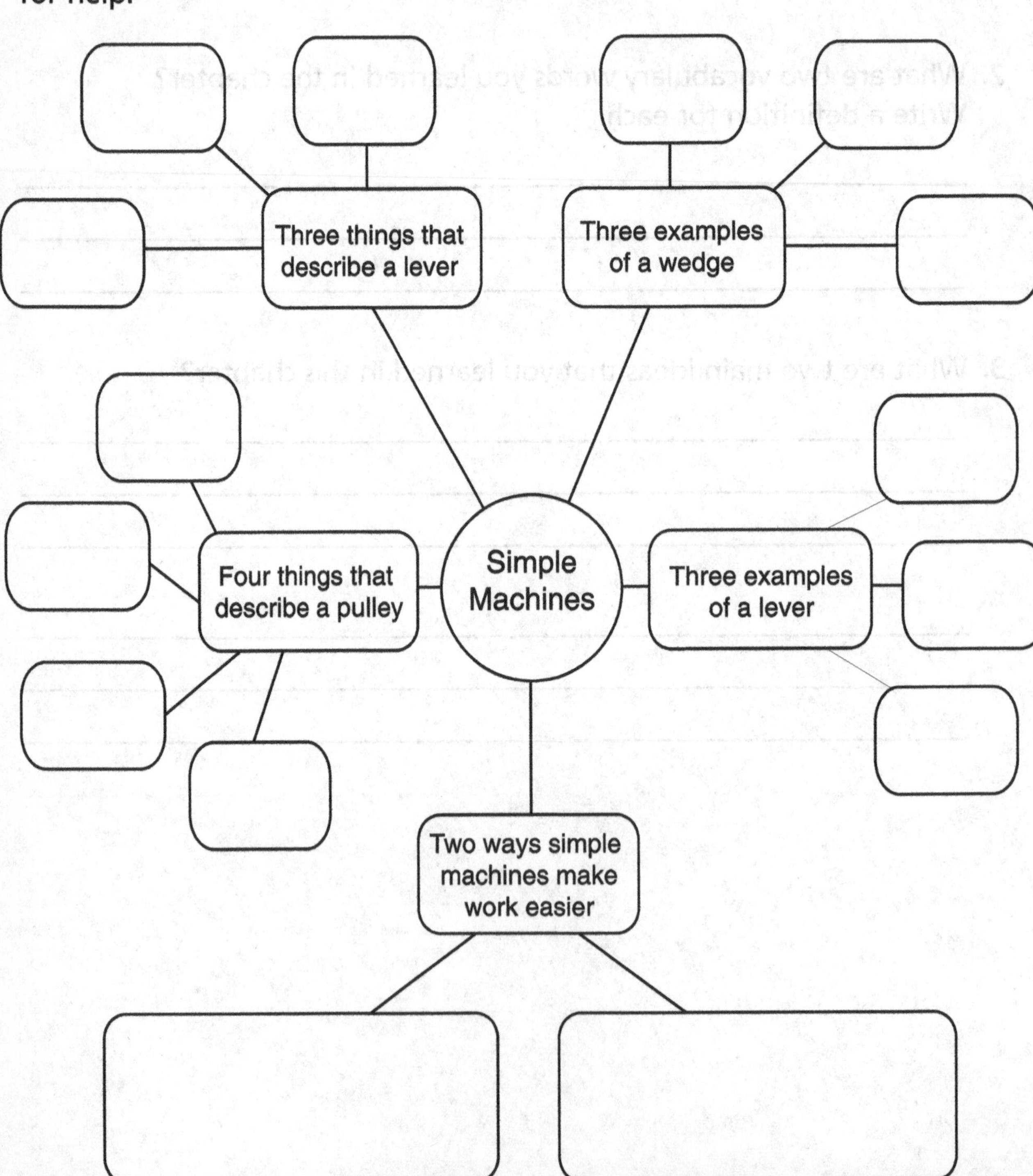

Name ________________________ Date ____________

Chapter Reading Skill
Chapter 10

Compare and Contrast

When you **compare,** you tell how things are alike. When you **contrast,** you tell how things are different. **Read the story. Use the chart to compare and contrast the two characters in the story.**

A long time ago, when our country was very young, two sisters lived on a farm. Marge and her sister Linda both liked to feed the cows, but they had different jobs. Everyday Marge would go to the well to get the cows some water. The well had a pulley with a bucket attached to it. Marge would lower the bucket into the well to get the water. Linda would go to the field to get some hay. She would put the hay into a wheel barrow and bring it to the cows. Linda's favorite cow was Nellie, a brown cow with spots. Marge liked Bessie the best. Bessie was a brown cow, but she did not have any spots. Every night Marge and Linda would say goodnight to the cows before they went to bed.

Make an X in the column to compare the two sisters.

	Marge	Linda
Lives on a farm		
Feeds the cows		
Gets water for the cows		
Gets hay for the cows		
Uses a pulley		
Uses a wheel barrow		
Likes Bessie the best		
Likes Nellie the best		
Says good night to the cows		

Name ______________________ Date ____________

Chapter Reading Skill
Chapter 10

Cars and Wagons

Sometimes, when you compare and contrast two things, it helps to make a diagram. **Use the diagram below to compare and contrast wagons and automobiles.**

Many years ago, before the car was invented, people traveled in wagons. Wagons cannot move under their own power as cars do. Wagons need to have another source of power. Many people used horses or mules to pull their wagons. Cars, on the other hand, are powered by gasoline. Even though cars and wagons are very different from each other, they are alike in many ways. Both cars and wagons can move forward because they have wheels. Without wheels, it would be very hard for cars and wagons to move. People use both cars and wagons to get from one place to another. Many people can ride inside, and people can carry things in them too. Even though both cars and wagons travel from place to place, cars travel much faster. Since cars were invented, it has become much easier for people to go places.

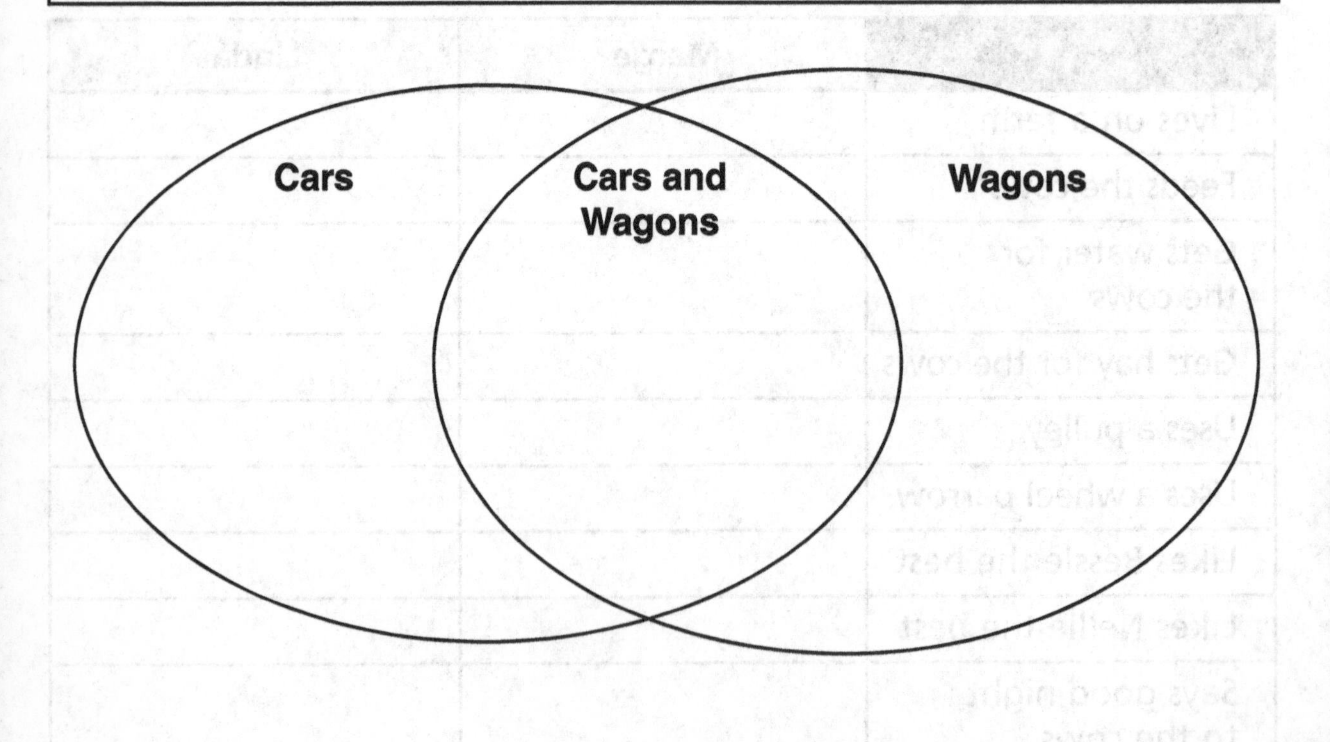

Name ______________________ Date __________

Lesson Outline
Lesson 4

Doing Work

Fill in the blanks. Reading Skill: **Compare and Contrast** - questions 14, 15, 16, 17

How Are Work and Energy Related?

1. When a force changes the motion of an object, ______________ is done.
2. When you pick up books, work is done because a(n) ______________ changes the motion of the books.
3. When you push against a wall, work is not done because there is no change in the ______________ of the wall.
4. Energy is needed to do ______________.
5. Energy exists in different ______________.
6. Moving things have energy of ______________.
7. Energy that can make an object move is called ______________ energy.
8. A rock rolling down a hill has ______________.
9. A rock on top of a hill has ______________ energy.
10. Other forms of energy include heat, light, sound, and ______________.

Name______________________ Date____________

Fill in the blanks.

How Does Energy Change?

11. Energy can change from one ______________ to another.
12. Friction changing to ______________ is an example of energy changing from one form to another.
13. The diagram on page E40 shows how the energy of ______________ moves from one ball to another.
14. The diagram shows how one ball causes another ball's motion to ______________.
15. The diagram shows that the ball being hit now has ______________.

Name_________________________ Date_____________

Interpret Illustrations

Lesson 4

How Are Doing Work and Energy Related?

Pictures often illustrate a concept. The pictures below show a student doing various activities.

Look at each picture and describe which of them show work being done.

1. What is the student doing in the first picture? ____________________

2. Is work being done in the first picture? Explain.

   ____

3. Is work being done in the second picture? Explain.

   ____

4. Which of the remaining two pictures shows work being done? Explain.

   ____

   ____

5. Which of the remaining two pictures shows no work being done? Explain.

   ____

   ____

Name_________________________ Date____________

Interpret Illustrations

Lesson 4

How Does Energy Change?

A diagram uses pictures and words to describe a thing or a process. The diagram below shows how the energy of motion can move from one object to another.

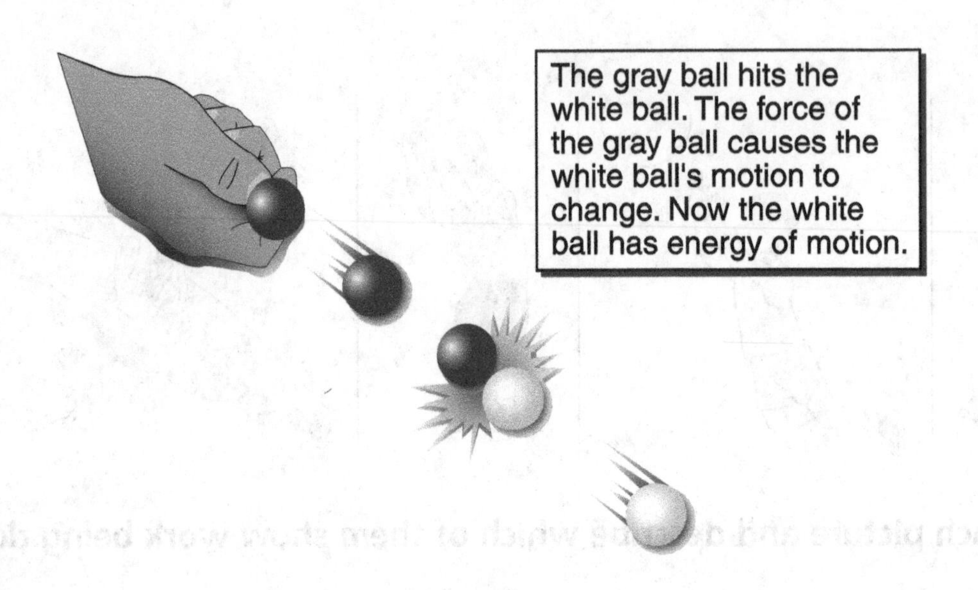

Answer these questions about the diagram above.

1. Which of the balls has the energy of motion?

____

2. What causes the white ball's motion to change?

____

3. What happens to the white ball when it is hit by the gray ball?

____

____

4. Why does the white ball move?

____

____

Name____________________________ Date______________

Lesson Vocabulary
Lesson 4

Doing Work

Fill in the blanks. You can use a word more than once.

Vocabulary
form
stored
force
forms
energy
motion
move
work
batteries

1. When you pedal a bicycle, you are doing ____________________.
2. Work is done when a(n) ____________________ changes the motion of an object.
3. The ability to do work is called ____________________.
4. Energy exists in different ____________________.
5. Moving things have energy of ____________________.
6. Energy that can make an object move is called ____________________ energy.
7. Food, fuel, and ____________________ are all forms of stored energy.
8. Heat, light, sound, and electricity are all forms of ____________________.
9. Energy can ____________________ from one object to another.
10. Energy can change from one ____________________ to another.

Name________________________________ Date______________

Cloze Test

Lesson 4

Doing Work

Vocabulary

force	energy	stored
motion	form	

Fill in the blanks.

Whenever a(n) ____________________ changes the motion of an object, work is done. The ability to do work is ____________________. Moving things, such as balls rolling, have energy of ____________________. Energy that can make an object move is called ____________________ energy. Energy changes from one ____________________ to another.

Name________________________________ Date______________

Lesson Outline

Lesson 5

Levers and Pulleys

Fill in the blanks. Reading Skill: **Compare and Contrast** - questions 4,19

How Can You Make Work Easier?

1. A machine is a tool that makes ________________ easier to do.
2. A machine can change the ________________ of the force needed to do work.
3. A machine can change the ________________ of force needed to do work.
4. Some machines can change both the direction and the amount of ________________ needed to do work.
5. Machines with few or no moving parts are called ________________ machines.
6. A straight bar that moves on a fixed point is a(n) ________________.
7. All levers have three important parts—the load, the fulcrum, and the ________________.
8. A lever lets you ________________ the direction of a force.
9. A lever lets you change the ________________ of force needed to move something.
10. The force is the push or pull that ________________ the lever.
11. The object being lifted or moved by a lever is called the ________________.
12. The point where a lever turns is the ________________.

Name______________________________ Date____________

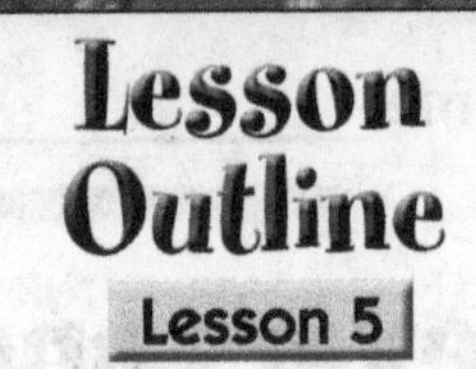

Fill in the blanks.

Are There Different Kinds of Levers?

13. The force, ____________________, and load of a lever can change places.

14. The load is located in the ____________________ of a wheelbarrow.

What Are Some Other Simple Machines?

15. A wheel and axle is a simple machine with a(n) ____________________ that turns on a post.

16. A wheel and axle makes work easier by changing the ____________________ of the turning force.

17. When you turn the handle of a windlass in a large circle, the ____________________ turns in a small circle.

What Goes Down to Go Up?

18. A simple machine that uses a wheel and a rope to lift a load is a(n) ____________________.

19. Some pulleys make work easier by reducing the amount of force needed to ____________________ the load.

Name________________________________ Date______________

Interpret Illustrations

Lesson 5

How Levers Work

A diagram uses pictures and words to describe a thing or process. This diagram shows a lever. The words in the boxes describe the parts of a lever.

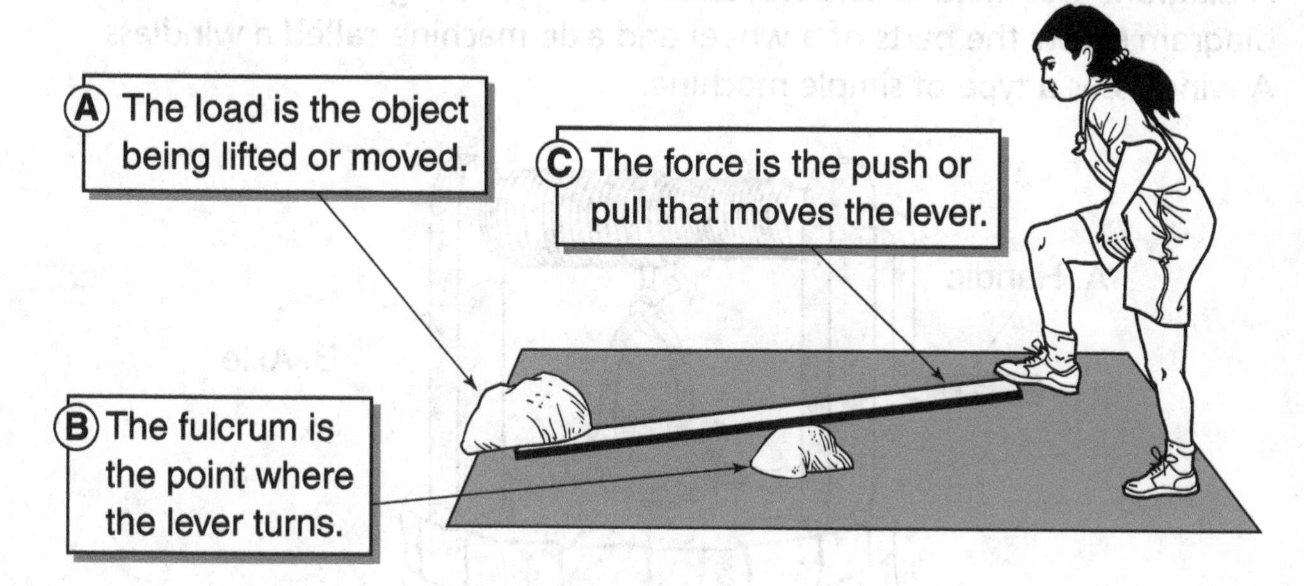

Use the diagram to answer the questions.

1. What part of a lever is shown in Part A of the diagram?

   ____

2. What part of a lever is shown in Part B of the diagram?

   ____

3. What part of a lever is shown in Part C of the diagram?

   ____

4. What part is the point where the lever turns? ______________________

5. Which part of the diagram shows the object being lifted or moved?

   ____

6. Which part of the diagram shows a push or a pull?

   ____

7. What are two ways that using a lever makes it easier to do work?

   ____

   ____

Name_________________________ Date_____________

Interpret Illustrations
Lesson 5

What Are Some Other Simple Machines?

A diagram uses pictures and words to describe a thing or process. This diagram shows the parts of a wheel and axle machine called a windlass. A windlass is a type of simple machine.

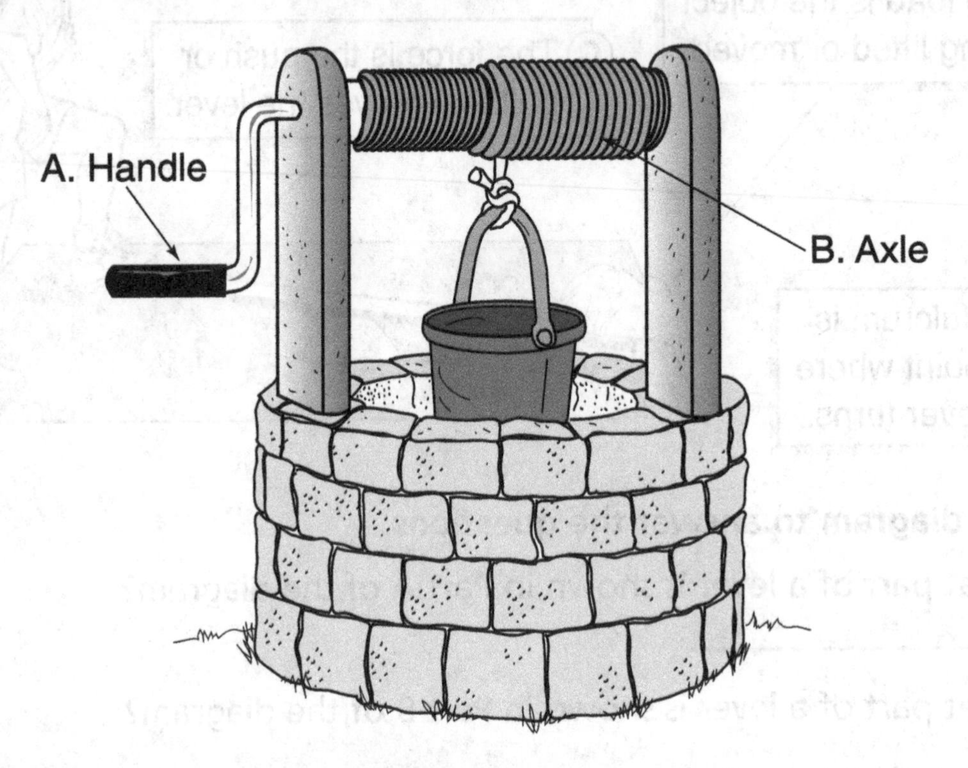

Use the diagram to answer the questions.

1. What part of a windlass is shown in Part A of the diagram? __________
2. What part of a windlass is shown in Part B of the diagram? __________
3. What happens to the axle when the handle is turned? __________
4. Which turns in a larger circle, the handle or the axle? __________
5. What happens to the bucket when the axle turns?
____
6. What is the job of the rope? __________
____

Name____________________ Date__________

Lesson Vocabulary
Lesson 5

Levers and Pulleys

Vocabulary

amount
fulcrum
pulley
machine
lever
strength
force
simple
wheel and axle
levers

Fill in the blanks.

1. A tool that makes work easier to do is a(n) ____________________.
2. A machine can change the direction of the ____________________ needed to do work.
3. A machine can change the____________________ of force needed to do work.
4. A machine with few or no moving parts is a(n) ____________________ machine.
5. A(n) ____________________ is a straight bar that moves on a fixed point.
6. The point where a lever turns is the ____________________.
7. A simple machine that has a wheel that turns on a post is called a(n) ______________________________.
8. A wheel and axle makes work easier by changing the ____________________ of the turning force.
9. A simple machine that uses a wheel and a rope to lift a load is a(n) ____________________.
10. Both a pulley and a wheel and axle are types of ____________________.

Name ______________________ Date ______________

Cloze Test
Lesson 5

Levers and Pulleys

Vocabulary

simple	direction	load
machine	fulcrum	

Fill in the blanks.

You can make work easier by using a tool called a(n) ______________.
A lever is one example of a(n) ______________ machine. A lever allows you to change the ______________ and amount of a force needed to move a load. The point where the lever turns is the ______________. The ______________ is the object being lifted or moved by the lever.

Name_______________________ Date___________

Lesson Outline
Lesson 6

More Simple Machines

Fill in the blanks.  Reading Skill: **Compare and Contrast** - questions 16,17,18

What Is an Inclined Plane?

1. A ramp is a(n) ________________ that is higher at one end.
2. A ramp is also called a(n) ________________.
3. An inclined plane makes work easier by moving an object over a longer distance with less ________________.
4. The people who built the Pyramid of the Sun may have used an inclined plane to move ________________.
5. Two inclined planes placed back to back form a(n) ________________.
6. Some wedges use ________________ to raise an object.
7. Some wedges use force to ________________ objects apart.
8. A wedge used to split wood apart is the ________________.
9. The downward force of an ax changes to the ________________ force that splits the wood.
10. A wedge used by farmers is a(n) ________________.

Name ______________________ Date __________

Lesson Outline
Lesson 6

Fill in the blanks.

What Is a Screw?

11. A screw is a(n) ______________ wrapped into a spiral.

12. The ridges of a screw are called ______________.

13. A screw with a(n) ______________ inclined plane has more threads.

14. A screw with a(n) ______________ inclined plane has fewer threads.

What Happens If You Put Two Simple Machines Together?

15. Two or more simple machines put together form a(n) ______________.

16. A pair of scissors is a compound machine. Part is a lever. Part is a(n) ______________.

17. A water faucet is a compound machine. Part is a screw. Part is a(n) ______________.

18. A bicycle uses wheels and axles, and a(n) ______________.

Name________________________________ Date____________

Interpret Illustrations

Lesson 6

What is an Inclined Plane?

Diagrams often tell us how something works. This diagram shows how a simple machine makes work easier.

Inclined plane:
longer distance, less effort

Straight up:
shorter distance,
more effort

Use the diagram to answer the questions below.

1. What simple machine is shown in the diagram?

____

2. Which of the paths shown in the diagram is shorter?

____

3. Which of the paths shown in the diagram takes less effort?

____

____

4. You have a very heavy object to move. Which path in the diagram would you choose? Explain.

____

____

Name_____________________________ Date_____________

Interpret Illustrations

Lesson 6

What is a Screw?

Diagrams tell us how something is put together. The diagram below shows how a screw is formed.

Diagram 1

A screw with a longer inclined plane has more threads. A screw with a shorter inclined plane has fewer threads.

Diagram 2

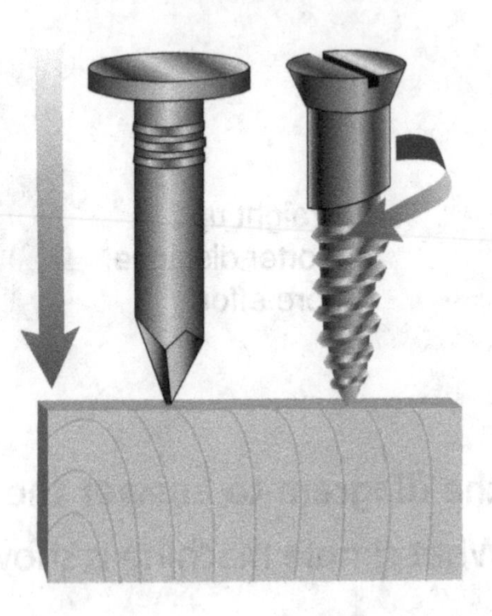

Answer these questions about the diagram above.

1. Which screw in Diagram 1 has more threads?

____

2. Which screw in Diagram 1 has a shorter inclined plane? How do you know?

____

____

3. In Diagram 2, which one is the screw? Tell how you know.

____

____

Name____________________ Date__________

Lesson Vocabulary
Lesson 6

More Simple Machines

Fill in the blanks. You may use a word more than once.

Vocabulary
incline plane
compound
distance
ramp
force
screw
wedge
lever

1. A flat surface that is higher at one end is a(n) ________________.
2. A ramp is also called an ________________.
3. A(n) ________________ is a simple machine made up of two inclined planes placed back to back.
4. A wedge uses ________________ to raise an object or split objects apart.
5. An inclined plane wrapped into a spiral forms a(n) ________________.
6. A screw allows you to apply force over a longer ________________.
7. Two or more simple machines put together form a(n) ________________ machine.
8. A pair of scissors is a compound machine that has a part that is a lever and a part that is a(n) ________________.
9. A bicycle uses wheels and axles, and a(n) ________________.

Name ______________________ Date ____________

Cloze Test
Lesson 6

More Simple Machines

Vocabulary

force	**threads**	**inclined plane**
distance	**screw**	

Fill in the blanks.

A(n) ________________ wrapped into a spiral is a screw. This simple machine has ridges called ________________. You need less force to turn a(n) ________________ than to pound a nail. This is because you are applying force over a longer ________________ when you use a screw. The longer the distance, the less ________________ you need to do work.

Name______________________________ Date______________

Chapter Vocabulary

Chapter 10

Work and Machines

Circle the letter of the best answer.

1. Work is done when a force changes the
   a. distance of an object.
   b. motion of an object.
   c. height of an object.
   d. weight of an object.

2. In order to do work, you need
   a. energy.
   b. gravity.
   c. distance.
   d. weight.

3. A ball rolling down a hill has energy of
   a. direction.
   b. location.
   c. motion.
   d. storage.

4. A tool that makes work easier to do is a
   a. force.
   b. fulcrum.
   c. machine.
   d. map.

5. Machines with few or no moving parts are
   a. bicycles
   b. faucets.
   c. loads.
   d. simple machines.

6. The point where a lever turns is the
   a. axle.
   b. force.
   c. fulcrum.
   d. load.

Name ______________________ Date ____________

Chapter Vocabulary

Chapter 10

Circle the letter of the best answer.

7. A wheel and axle has a wheel that turns on a
   a. fulcrum. b. lever.
   c. load. d. post.

8. A simple machine that uses a wheel and a rope to lift a load is a(n)
   a. axle. b. lever.
   c. pulley. d. windlass.

9. All of the following are types of levers EXCEPT a
   a. pulley. b. ramp.
   c. seesaw. d. wheel and axle.

10. A flat surface that is higher at one end is a ramp. A ramp is also called a(n)
   a. inclined plane. b. lever.
   c. pulley. d. wheel and axle.

11. An example of a wedge is a(n)
   a. ax. b. bicycle.
   c. water faucet. d. wheelbarrow.

12. An inclined plane wrapped into a spiral forms a
   a. lever. b. ramp.
   c. screw. d. wedge.

Crack-a-Code

Use the Code Key to help you decode each word. Then draw a line to its meaning.

Code Key

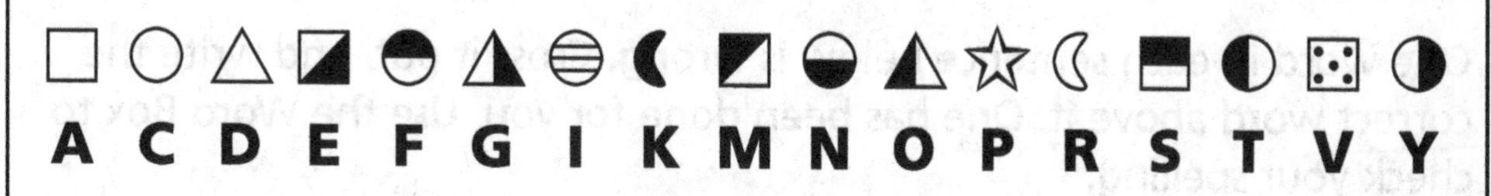

1. ______________________ a. force when one object rubs against another

2. ______________________ b. needed to be able to work

3. ______________________ c. pulling force between objects

4. ______________________ d. change of position

5. ______________________ e. the space between two objects

6. ______________________ f. how fast something moves

7. ______________________ g. all pushes and pulls

Name ______________________ Date ____________

Unit Vocabulary
Unit E

Correct-a-Word

Word Box				
newton	friction	motion	pulley	weight
location	simple	lever	inclined	plane

One word in each sentence below is wrong. Cross it out, and write the correct word above it. One has been done for you. Use the Word Box to check your spelling.

1. A machine with few or no moving parts is a ~~single~~ simple machine.
2. A screwdriver is a kind of liver.
3. A pushy uses a wheel and a rope to lift a load.
4. Position is the locate of an object.
5. Fraction is a force that occurs when things rub together.
6. The unit used to measure force is a neutron.
7. Work happens when a force changes an object's matter.
8. A flat surface raised at one end is called an inclined plan.
9. The pound is used to measure force and friction in the English system.

Name ______________________ Date ______________

Write-a-Word

gravity	machine	data	weight	work
pulley	speed	friction	location	force

Choose a word from the box above to complete each sentence below. Write the word in the sentence.

1. What's the ______________ limit on this highway?
2. We gather lots of ______________ for our report.
3. The force of ______________ pulls a plant's roots down into the ground.
4. That's the exact ______________ where we found the treasure chest.
5. I wish I had a ______________ that made my homework easier!
6. We put a rope over a wheel to make a ______________ that helped us lift the heavy box.
7. The pull of gravity on an object determines the object's ______________.
8. When a force changes the motion of an object, ______________ is done.
9. When you rub your hands together, you feel heat from ______________.
10. You use ______________ when you give an object a push or a pull.

Name ______________________ Date ____________

Chapter Graphic Organizer
Chapter 11

Matter

Matter can be a solid, a liquid, or a gas. Each form of matter has characteristics that make it different from the other forms of matter. Look at the chart below, then answer the questions that follow it.

Substance	Form of Matter	Arrangement of Particles
lead	solid	very close together; form a pattern
water	liquid	close together; do not form a pattern
calcium	solid	very close together; form a pattern
oxygen	gas	very far apart
milk	liquid	close together; do not form a pattern
nitrogen	gas	very far apart

1. What characteristics do liquids share?

____

____

2. What characteristics do solids share?

____

____

3. What characteristic do gases share?

____

____

Name______________________________ Date______________

Chapter Reading Skill
Chapter 11

Draw Conclusions

A conclusion is a summing-up statement that is based on facts. You can draw conclusions about what you see in a picture. Or you can draw conclusions about what you read. Conclusions may be about people, places, or events.

Look at the picture below. Draw a conclusion about what happened.

1. Write what you know from looking at the picture.

____

____

2. Write what you already know about cats.

____

____

3. Draw a conclusion about what happened.

____

____

____

Name ______________________ Date __________

Chapter Reading Skill

Chapter 11

The Snowman

In science, we draw conclusions from facts and from the results of experiments. We can draw conclusions by making observations and from recording data. You can also draw conclusions from what you read.

Read the paragraph, then answer the questions below.

One cold, winter night, it began to snow. It snowed all night long. When Jenna and Pete woke up, there were three feet of snow on the ground.

"Let's make a snowman," cried Pete.

Quickly, the children put on warm clothes, gloves, and scarves. Jenna looked at the thermometer. It was 25°F outside.

"Wow, it sure is cold," she said.

Once outside, Jenna and Pete built a snowman, rolling small snowballs until they were very large. They stacked up three large snowballs to make the snowman, then they pushed rocks into the snow to form eyes, a nose, and a mouth.

The next day, Jenna and Pete went to school. It became very warm that day. The temperature had gone up to 35°F outside. When the children came home from school, they were surprised to see that their snowman was starting to melt.

"What happened?" cried Jenna.

"Let's find out," said Pete.

Clues from the text:

____

____

What I know:

____

My conclusion:

____

Name________________________________ Date______________

Lesson Outline
Lesson 1

Properties of Matter

Fill in the blanks.  Reading Skill: **Draw Conclusions** - questions 1, 11, 19

How Do We Describe Matter?

1. Since a book and a pencil both take up space, they are both types of ______________.
2. Matter can be described by its ______________.
3. A property is any characteristic of matter that you can ______________.
4. Size, color, and shape are ______________.
5. A property of a dime is ______________.
6. The more space an object takes up, the greater its ______________.
7. Since a tennis ball takes up less space than a bowling ball, a tennis ball has less ______________ than a bowling ball.
8. The amount of matter that is in an object is called its ______________.
9. The mass of a beach ball is ______________ than the mass of a bowling ball.

Name________________________________ Date______________

Lesson Outline
Lesson 1

Fill in the blanks.

How Do You Measure Mass and Volume?

10. Matter is made of very tiny ______________.
11. Since the particles in a bowling ball are packed more tightly than the particles in a balloon, a bowling ball has more ______________ than a balloon.
12. Since a paper clip is small, its mass is measured in ______________.
13. Since a school bus is large, its mass is measured in ______________.
14. One unit used to measure ______________ is the liter.

How Are Mass and Weight Related?

15. The greater the mass of an object, the greater its ______________.
16. The weight of an object depends on the pull of ______________ on that object.
17. Since the Moon has less ______________ than Earth, the pull of gravity is weaker on the Moon than on Earth.
18. An example of a force is ______________.
19. Forces can be measured in units called ______________.

Name________________________________ Date______________

Interpret Illustrations

Lesson 1

How Do We Describe Matter?

A chart is helpful when you want to organize information. This chart shows the physical properties of some familiar objects.

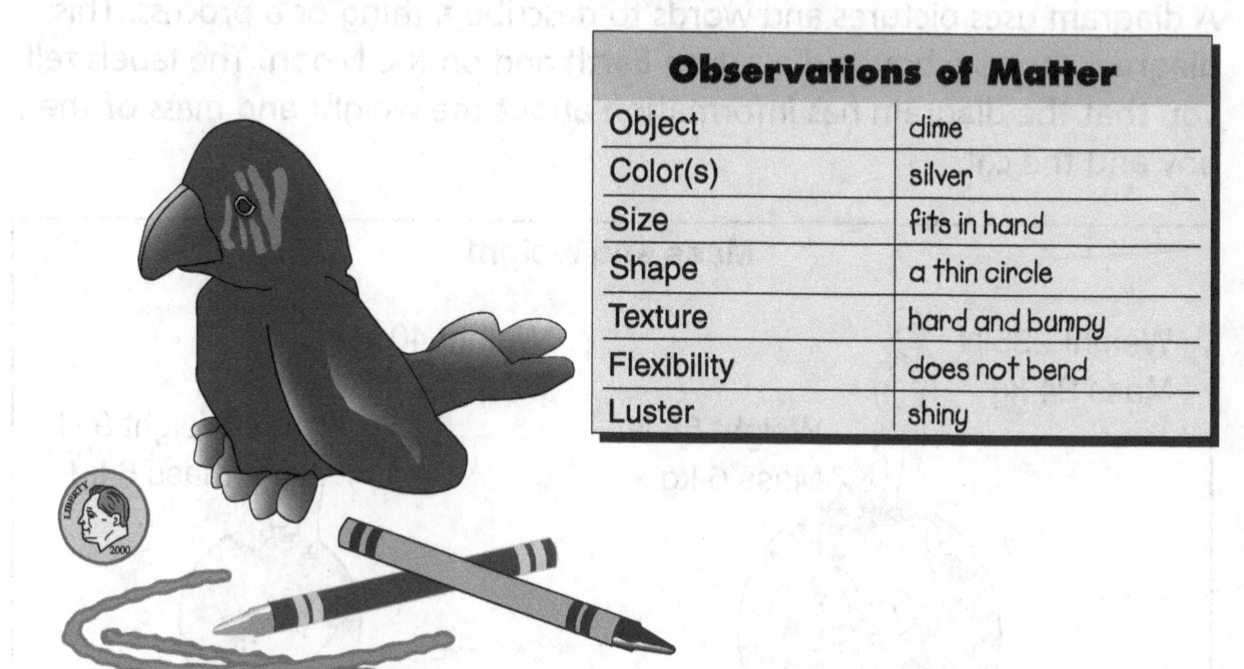

Observations of Matter	
Object	dime
Color(s)	silver
Size	fits in hand
Shape	a thin circle
Texture	hard and bumpy
Flexibility	does not bend
Luster	shiny

Use the chart to answer the following questions.

1. What is the information in the chart about? ________________
2. What is the texture of the dime? ________________
3. What is the luster of the dime? ________________
4. What is the size of the dime? ________________
5. Look at the illustration of the other objects. What could you say about the flexibility of the pipe cleaners? ________________
6. How would you describe the shape of the crayons?

____

Name________________________ Date____________

Interpret Illustrations
Lesson 1

How Are Mass and Weight Related?

A diagram uses pictures and words to describe a thing or a process. This diagram shows a boy and a cat on Earth and on the Moon. The labels tell you that the diagram has information about the weight and mass of the boy and the cat.

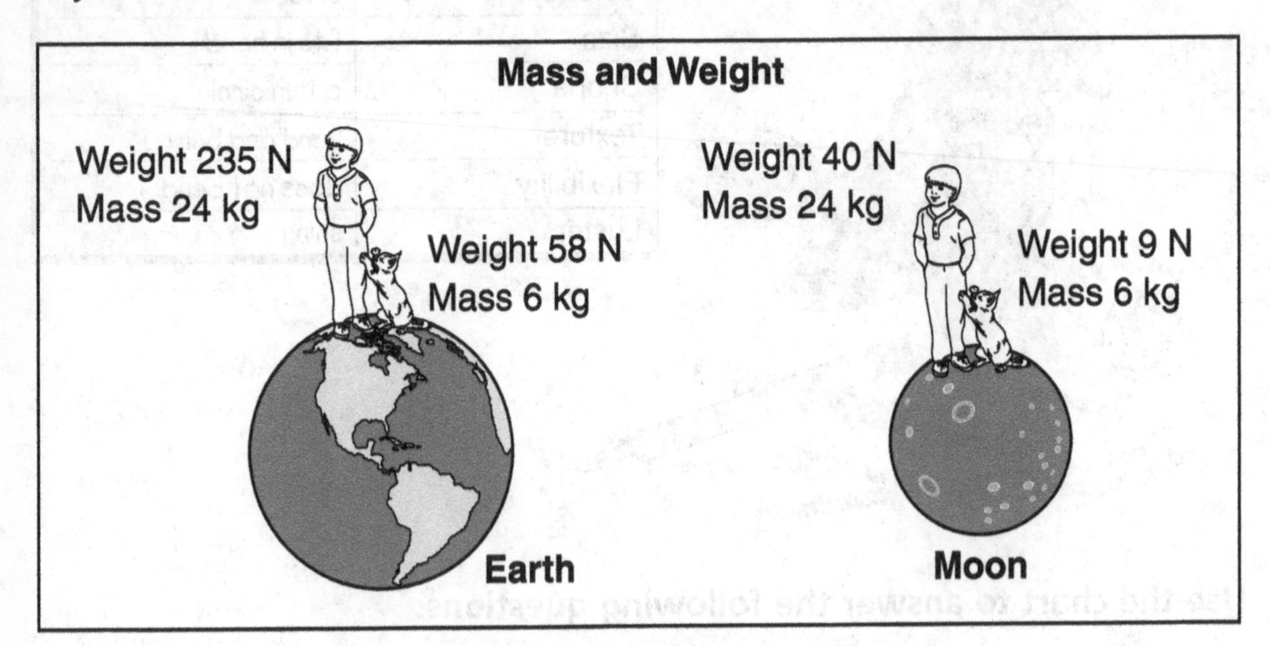

Use the diagram to answer the following questions.

1. What is the weight of the boy on Earth? On the Moon?____________
2. What is the mass of the boy on Earth? On the Moon?____________
3. What is the weight of the cat on Earth? On the Moon?____________
4. What is the mass of the cat on Earth? On the Moon?____________
5. Look at the weight of the boy and the weight of the cat in each location. In which location are the weights less? In which are they more?
____
6. What do you notice about the mass of the boy and the mass of the cat in each location?____________________________

Name________________________________ Date____________

Lesson Vocabulary
Lesson 1

Properties of Matter

Use one of these words to complete sentences 1–6.

Vocabulary
gravity
space
properties
matter
mass
volume

1. The amount of space an object takes up is called its ______________.
2. Mass is a measure of how much ______________ an object contains.
3. The weight of an object is the pull of ______________ on the object.
4. An object with a large ______________ feels heavy.
5. Matter is anything that takes up ______________ and has mass.
6. You can describe matter by naming its ______________.

Answer these questions in your own words.

7. You have one bag of trash with a mass of 3 kg and another bag of trash with a mass of 14 kg. What do you know about the weights of the bags of trash?

____

____

8. You have a bowl of pancake batter. You add a quart of blueberries to the batter. What happens to the level of the batter in the bowl when you add the blueberries?

____

9. A dog weighs 70 N on Earth. If the dog is sent to the Moon, how would its weight change? How would its mass change?

____

____

Name________________________________ Date______________

Cloze Test
Lesson 1

Properties of Matter

Vocabulary

property	texture	mass
volume	sink	

Fill in the blanks.

You can describe an object by naming a(n) ________________, or characteristic, of the object. All matter has ________________ and ________________. Other properties are special for each type of matter, such as size, shape, color, and ________________. There are more properties you can use to describe matter. For example, some objects float in water and other objects ________________.

Name________________________ Date____________

Lesson Outline
Lesson 2

Comparing Solids, Liquids, and Gases

Fill in the blanks. Reading Skill: **Draw Conclusions** - question 5, 14

How Can You Classify Matter?

1. Solids, liquids, and gases are alike in that they all take up space and have ________________.
2. All solids have a definite shape and ________________.
3. All liquids have a(n) ________________ volume.
4. A liquid does not have a definite ________________.
5. Since milk takes the shape of the container it is in, it is a(n) ________________.
6. A gas has no definite shape or ________________.
7. Both a gas and a(n) ________________ take the shape of the container they are in.
8. All solids, liquids, and gases are made of ________________.
9. The particles in a solid form a certain ________________.
10. The particles in a liquid have more ________________ than the particles in a solid.
11. The particles in a gas have more energy than the particles in a solid or a ________________.
12. The particles in a(n) ________________ can spread out to fill a large container or squeeze together to fit a small container.

Name______________________________ Date____________

Fill in the blanks.

How Can Matter Change?

13. Matter can change form and still be the same type of ________________.

14. A change in how matter looks, but not in the kind of matter it is, is called a(n) ________________.

15. Ice becomes liquid water when it ________________.

16. Water that evaporates changes from a liquid to a(n) ________________.

17. When water vapor cools it loses energy and ________________.

Can You Mix Different Kinds of Matter Together?

18. A mixture is a combination of different forms of ________________.

19. The properties of each type of matter in a(n) ________________ do not change.

What Is a Solution?

20. When one or more types of matter are mixed evenly in another type of matter, a ________________ is formed.

21. Salt water is a type of ________________.

Name ______________________ Date __________

Interpret Illustrations

Lesson 2

How Can You Classify Matter?

A diagram uses pictures and words to describe a thing or a process. This diagram shows particles in different states of matter.

Particles in Different States of Matter

Solid

Liquid

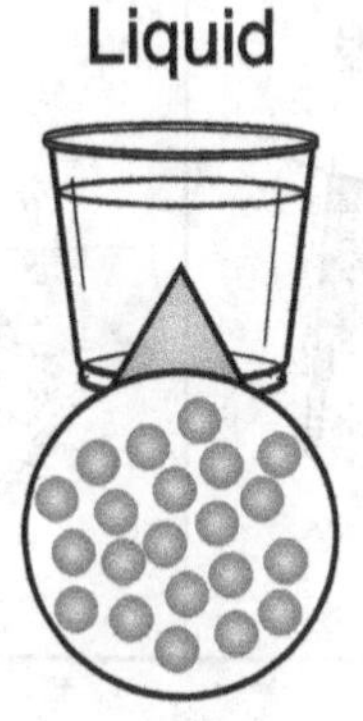

Gas

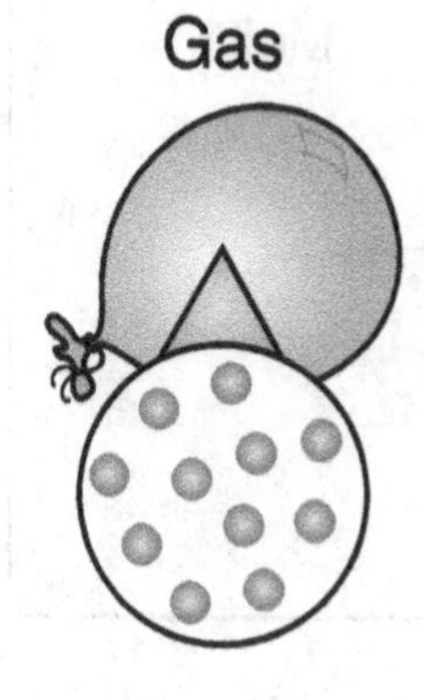

Use the diagram to answer the following questions.

1. What is the title of this diagram?

____

2. What states of matter are shown in the diagram?

____

3. What do the close-ups show?

____

____

4. In which state of matter are the particles packed tightly together?

____

5. Which state of matter has the most space between its particles?

____

6. Describe the particles that make up a liquid.

____

Name________________________________ Date______________

Interpret Illustrations

Lesson 2

How Can Matter Change?

Some charts use labels and pictures to describe processes. This chart shows how the three states of matter change from one to another.

Changes of State

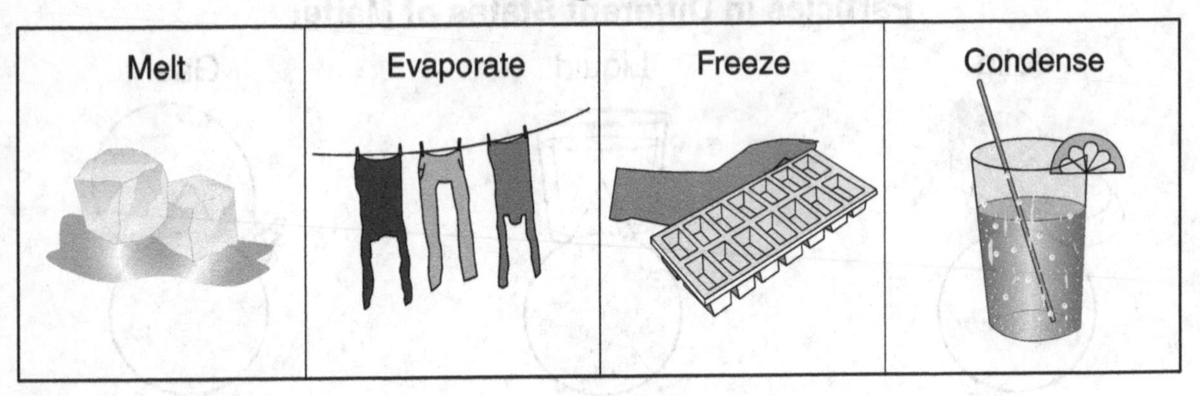

Use the chart to answer the following questions.

1. What three states of matter are shown in the chart?

____

2. What are the four ways that matter can change?

____

3. What happens when water is changed to ice cubes?

____

4. What happens to water when wet clothes dry on a clothesline?

____

5. What happens to water vapor when it cools on a glass of ice tea and becomes a liquid?

____

6. Describe what happens when ice melts.

____

Name______________________________ Date____________

Lesson Vocabulary
Lesson 2

Comparing Solids, Liquids, and Gases

Vocabulary

solid
physical change
mixture
liquid
gas
solution
shape

Use one of these words to complete sentences 1–7.

1. Solids have a definite ________________.
2. A ________________ is a change in how matter looks, but not in the kind of matter it is.
3. The state of matter that has no definite shape or volume is ________________.
4. Ice is the ________________ form of water.
5. Fruit salad is a(n) ________________.
6. Matter that has a definite volume, but does not have a definite shape is ________________.
7. Lemonade is a(n) ________________.

Answer these questions in your own words.

8. Can all three states of matter exist in a mixture? If so, give an example of such a mixture.

____

____

____

9. You can shape a certain type of material just as if it were clay. Left alone at room temperature, the material forms a puddle. What form of matter is the material? Why?

____

____

____

Name ______________________ Date __________

Cloze Test

Lesson 2

Comparing Solids, Liquids, and Gases

Vocabulary

vapor	solid	liquid
condenses	evaporates	

Fill in the blanks.

When water changes to a ______________ form, it is called ice. As ice melts, it changes from a solid into a(n) ______________. The particles in water have more energy than the particles in ice. If water is warmed, it ______________ and changes into a gas. This gas is called water ______________. When water vapor cools it loses energy and ______________.

Name______________________ Date__________

Lesson Outline
Lesson 3

Building Blocks of Matter

Fill in the blanks. Reading Skill: **Draw Conclusions** - questions 5, 8, 14, 17, 20

What Are Metals?

1. ______________ is a hard, shiny material found in Earth's ground.
2. Anything that attracts metals has the property of ______________.
3. A material that attracts iron and some other metals is called a ______________.
4. A strong metal that is used to make bridges, railroads, and cars is ______________.
5. Since aluminum is strong and light, it is used to make ______________.

What Are the Building Blocks of Matter?

6. Metals such as iron, gold, silver, and copper are called ______________.
7. The smallest particle of matter is called a (an) ______________.
8. A gas that fills a balloon causes the balloon to float, so the gas could be ______________.
9. Water is made up of the elements ______________ and ______________.
10. All of the elements are listed in the ______________.
11. The tusks of a walrus contain the element ______________.

Name________________________________ Date______________

Lesson Outline

Lesson 3

Fill in the blanks.

What Happens When Elements Join Together?

12. When two or more elements join together, it is called a ________________.

13. Compounds have very different ________________ from the elements of which they are made.

14. Since wood is made up of carbon, oxygen, and hydrogen, it must be a ________________.

15. The elements iron and oxygen form a compound called ________________.

16. Table salt is made up of the elements ________________ and ________________.

17. When you start with one kind of matter, and end up with another kind of matter, a ________________ has taken place.

18. An example of a chemical change is ____.

19. Burning wood breaks apart to form ________________ and ________________.

How Do Chemical and Physical Changes Compare?

20. Ice melting is an example of a ________________ change.

21. When an apple ripens, it turns red. This change of color is an example of a ________________ change.

What Are Metals?

A chart is a way to organize information. This chart contains information that tells about magnets.

What Is a Magnet?

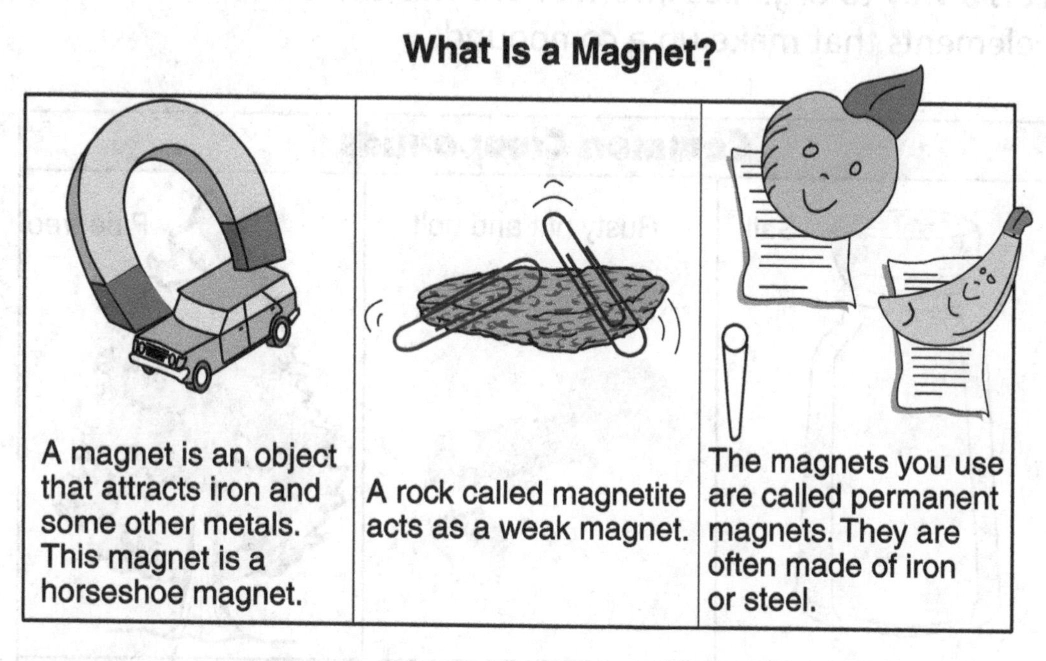

Use the chart to answer the following questions.

1. What is the title of the chart? ____________________
2. What kind of magnet is shown in the first column of the chart?

   ____

3. What kind of magnet is described in the third column of the chart?

   ____

4. What kind of rock acts as a weak magnet? ______________
5. What are permanent magnets made from? ______________
6. How can you tell if something is a magnet?

   ____

Name______________________ Date__________

Interpret Illustrations

Lesson 3

What Happens When Elements Join Together?

A chart is a way to organize information. This chart contains information about elements that make up a compound.

Common Compounds		
Salt	Rusty nut and bolt	Pine tree
Sodium + chlorine = salt	Iron + oxygen = rust	Carbon + oxygen + hydrogen = wood

Use the chart to answer the following questions.

1. What is the title of the chart? ____________________
2. What two elements are described in the first column of the chart?
   ____
3. What compound is shown in the first column? ____________
4. What do the + sign and the = sign in the first column tell you?
   ____
   ____
   ____

Name_________________________ Date___________

Lesson Vocabulary
Lesson 3

Building Blocks of Matter

Vocabulary

- iron
- elements
- atoms
- ground
- chemical change
- metal
- aluminum
- compound

Use one of the choices to complete sentences 1–8.

1. If an object is attracted to a magnet, it must be made of ______________.
2. Metals are found in the ______________.
3. Steel is made with ______________.
4. The building blocks of matter are called ______________.
5. All elements are made of ______________.
6. If the properties of elements change when they are mixed together, the mixture is called a(n) ______________.
7. One metal that is light and soft is ______________.
8. When a compound forms, a ______________ has taken place.

Answer these questions in your own words.

9. Which elements are found in water? How is the compound different from its elements?

____

____

____

10. How can magnets be used for work?

____

____

Name______________________________ Date____________

Cloze Test
Lesson 3

Building Blocks of Matter

Vocabulary

magnets	iron	magnetism
metal	ground	

Fill in the blanks.

Magnets attract objects made of some types of ____________. A metal is a shiny material that can be found in Earth's ____________. Magnets attract the metal called ________________. Anything that attracts metals has the property of ________________. The property of magnetism can be used to identify objects. Junkyards use powerful ______________ to sort certain metals from other objects.

Name________________________________ Date______________

Matter

Circle the letter of the best answer.

1. A gram is a unit that is used to measure
   a. length. b. speed.
   c. mass. d. weight.

2. Matter is
   a. anything that you can see.
   b. anything that takes up space and has mass.
   c. anything that you can touch.
   d. anything that has color.

3. Particles in a liquid have less energy than particles in
   a. water. b. a computer.
   c. ice. d. a gas.

4. Air is a mixture of different
   a. liquids. b. solids.
   c. gases. d. smells.

5. When one kind of matter is spread evenly throughout another type of matter, it is in a(n)
   a. solution. b. mixture.
   c. element. d. metal.

6. Everything in the world is made up of
   a. water. b. elements.
   c. vitamins and minerals. d. metals.

Name_________________________ Date___________

Chapter Vocabulary

Chapter 11

Circle the letter of the best answer.

7. Matter that has a definite volume but not a definite shape is called a(n)

a. solid. b. element.
c. liquid. d. gas.

8. Which of these is NOT a property of matter?

a. an opinion b. size
c. color d. shape

9. The amount of space that an object takes up is called its

a. size. b. shape.
c. weight. d. volume.

10. Blending different types of matter together makes a(n)

a. metal. b. mixture.
c. atom. d. element.

11. When different elements are blended to make a new material, such as sodium and chlorine combining to make salt, the result is called a

a. new element. b. metal.
c. compound. d. mixture.

12. Gold and silver are both

a. compounds. b. mixtures.
c. metals. d. liquids.

Chapter Summary

1. What is the name of the chapter you just finished reading?

____

____

2. What are two vocabulary words you learned in the chapter? Write a definition for each.

____

____

____

____

____

____

3. What are two main ideas that you learned in this chapter?

____

____

____

____

____

____

Name ______________________ Date ____________

Chapter Graphic Organizer
Chapter 12

Energy

Draw the correct arrows in the blank boxes. Each box should contain a light energy arrow or a heat energy arrow. Make sure you draw the arrows in the correct direction. After you are done, answer the questions that follow.

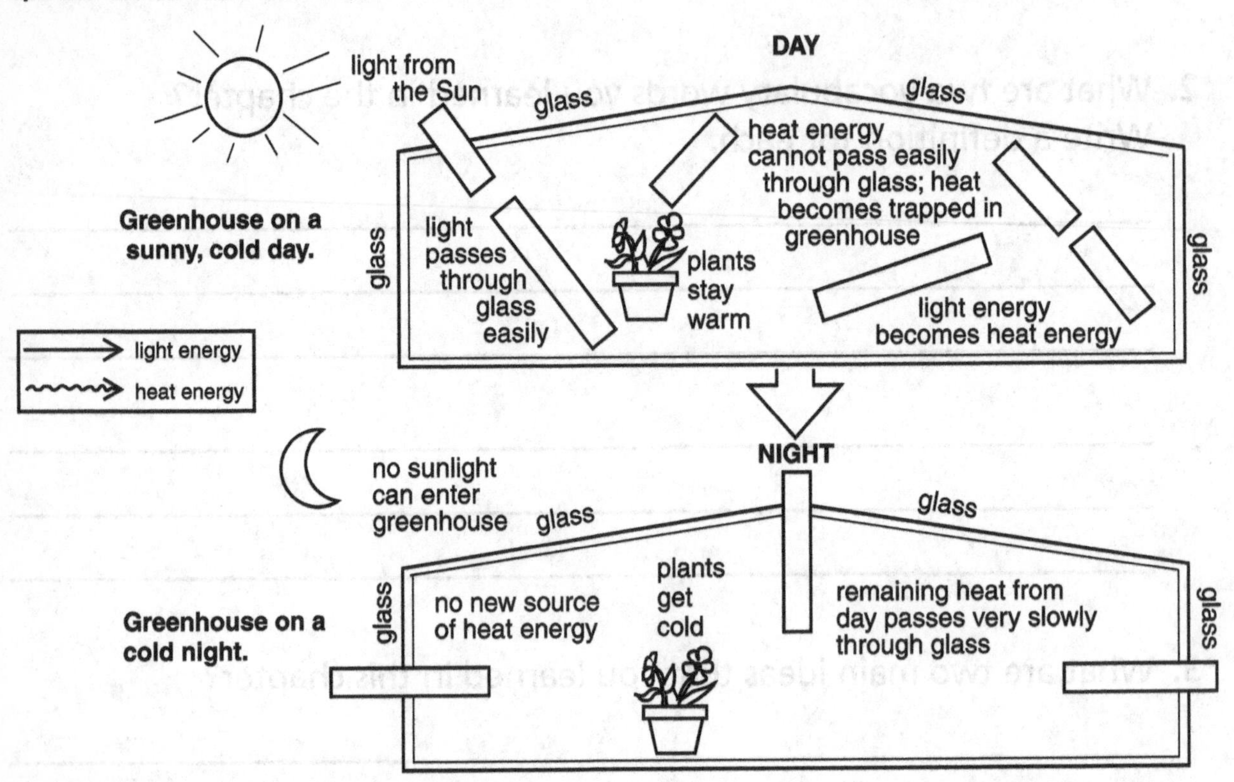

1. Would a heater be necessary in a greenhouse on a sunny but cold day? Why or why not?

____

____

____

2. Would a heater be necessary in a greenhouse on a cold night? Why or why not?

____

____

____

Name____________________________ Date______________

Chapter Reading Skill
Chapter 12

Sequence of Events

When you sequence events, you put them in the correct time order. Look for clue words such as *first, next, last, before,* and *after*. You can sequence the events that take place from start to finish.

Think about what you did this morning before you went to school. How did your day begin? What was the last thing you did before you left for school? Fill out the sequence chart below. Try to use the clue words above as you sequence the events that took place this morning.

Event 1

↓

Event 2

↓

Event 3

↓

Event 4

How to Make Water Chimes

Sometimes, you can sequence the events in a process or how something happens. Read the sentences below. Then use the clue words in the sentences to put the events in the correct time order. Write the numbers of the correct order in the spaces before each sentence.

_____ Continue marking the glasses, but each time increase the mark by one inch.

_____ Next, use the ruler to mark off 1 inch from the bottom of the first glass.

_____ First, you must gather the materials.

_____ After you have finished marking all 8 glasses, fill them with water to the level of each mark.

_____ Repeat the procedure with the next glass, but mark off 2 inches from the bottom.

_____ Finally, use the spoon to lightly tap the top of each glass. Each glass will make a different pitched sound.

_____ You will need 8 tall glasses, a marker, a ruler, and a spoon.

Play a tune with your glasses. Which glass will have the highest pitch? Explain.

____

____

____

Name ______________________ Date ____________

Lesson Outline
Lesson 4

How Heat Travels

Fill in the blanks. Reading Skill: **Sequence of Events** - questions 3, 4, 5

How Do Things Get Warmer?

1. ______________ is a form of energy that makes matter warmer.
2. Heat can move through solids, liquids, gases, and even ______________.
3. Heat ______________ from coils in a toaster into the bread to make toast.
4. The diagram on page F42 shows that heat moves from the flame to the pot to the ______________.
5. In the diagram, the temperature on the thermometer is ______________ on the second pot.
6. You can measure how much matter warms up by using a(n) ______________.
7. Thermometers measure ______________, which is how hot or cold an object is.
8. The unit of measurement for temperature is the ______________.

How Does Heat Change Matter?

9. Expand means to get ______________.
10. Heat causes the particles in ______________ to move faster.
11. When matter loses ______________, its particles slow down.
12. When matter contracts as it cools, it takes up ______________ space.
13. When a thermometer is in a warm place, the ______________ inside it expands and rises.

Name ______________________ Date ____________

How Can You Control the Flow of Heat?

14. Heat moves quickly through many ______________.

15. A(n) ______________ is a material that heat moves through easily.

16. An insulator is a material that ______________ doesn't travel through easily.

17. Wool, fur, and cotton are good ______________.

18. ______________ is an excellent insulator for animals such as the walrus.

19. Bears have very thick hair that helps trap ______________ inside the animal's body.

How Can Energy Change?

20. When fuel is ______________, heat is given off.

21. Your body uses ______________ to produce heat and keep the body at a certain temperature.

Name______________________ Date____________

Interpret Illustrations

Lesson 4

How Do Things Get Warmer?

Some diagrams use labels and pictures to describe a process. This diagram shows a before picture and an after picture to explain how heat travels.

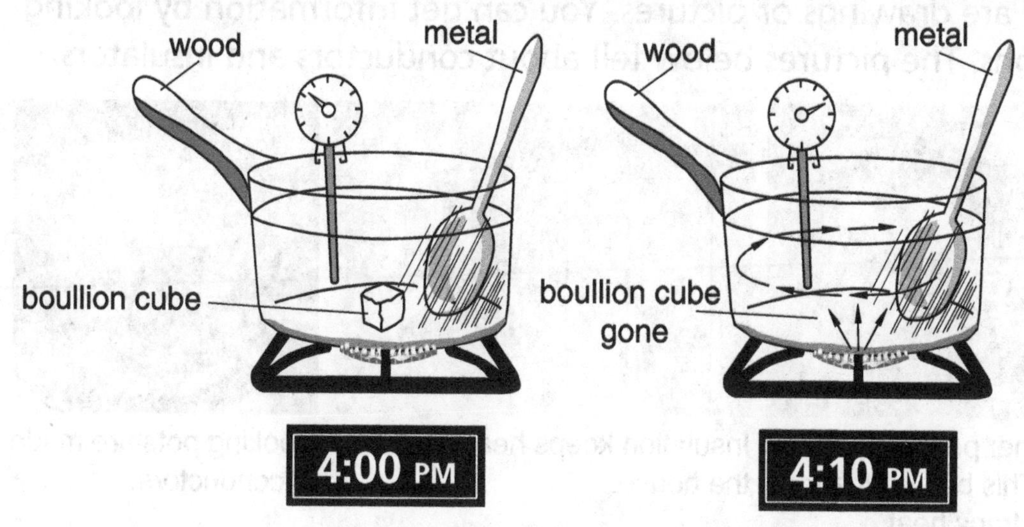

Heat moves from the flame to the pot to the soup. This is how the soup gets warmer.

Use the diagram to answer the questions.

1. Which of the thermometers shows a warmer temperature? How do you know?

____

____

2. Which of the two pots is warmer? How do you know?

____

____

3. How many minutes have passed?

____

4. What has happened to the cube?

____

5. In what directions does the heat travel?

____

Name_______________________________ Date_______________

Interpret Illustrations

Lesson 4

How Can You Control the Flow of Heat?

Illustrations are drawings or pictures. You can get information by looking at illustrations. The pictures below tell about conductors and insulators.

In cold weather people need to stay warm. This boy is wearing clothing that traps heat.

Insulation keeps heat inside the house.

Cooking pots are made of conductors.

Use the illustrations to answers the questions.

1. What does the first illustration show?

____

2. Does the picture of the boy show how a conductor or an insulator is used? Why?

____

3. How does the second picture show how an insulator is used?

____

____

4. What does the third illustration show?

____

5. Is the pot a conductor or an insulator? How do you know?

____

____

Name_______________________ Date___________

Lesson Vocabulary
Lesson 4

How Heat Travels

Use one of these words to complete sentences 1–7.

Vocabulary
thermometer
conductors
heat
temperature
degrees
insulators
expands

1. The ______________ of an object is a measure of how hot or cold that object is.
2. Matter ______________ when it is heated, and contracts when it is cooled.
3. ______________ is a form of energy that makes matter warmer.
4. You can use a(n) ______________ to measure temperature.
5. Metals are generally very good ______________ because heat flows through them quickly.
6. ______________ are materials that heat does not move through easily.
7. Temperature is measured in units called ______________.

Answer these questions in your own words.

8. You are going on a camping trip to the North Pole. What materials would you use in order to make your tent stay warm?

____

____

____

9. An egg is placed in a frying pan. The frying pan is then placed on the stove. How is heat transferred from the stove?

____

____

Name ______________________ Date ____________

Cloze Test

Lesson 4

How Heat Travels

Vocabulary

heat	hot	thermometer
degrees	energy	cold
temperature		

Fill in the blanks.

Temperature measures how ______________ or ______________ something is. Temperature can be measured by a(n) ______________. It is measured in ______________. The symbol for degree is °. A form of energy that makes matter warmer is ______________. Heat can be added to an object to raise the object's ______________. Some materials need more ______________ to raise their temperature than others.

Name____________________ Date__________

Lesson Outline
Lesson 5

How Light Travels

Fill in the blanks. Reading Skill: **Sequence of Events** - questions 4, 8, 9, 11, 12, 16

How Does Light Travel?

1. Light is a form of ________________.
2. Light can make ________________ move or change.
3. The Sun, lightning, and fire are all examples of ________________ light sources.
4. Light travels in ________________ from its source.
5. Materials that do not allow any light to pass through them are called ________________ materials.
6. Opaque materials create ________________.
7. When light hits an object, some light ________________ off the object.
8. You see an object because light reflected from the object enters your ________________.
9. Light rays bounce in many different directions from a dull or rough ________________.

Why Does Light Bend?

10. As light rays move from air to water or from water to air, they ________________.
11. Eyeglass lenses work by refracting ________________ into people's eyes.

Name ______________________ Date ____________

Lesson Outline

Lesson 5

Fill in the blanks.

Why Do You See Colors?

12. Light from the Sun is a mixture of many ______________.
13. When light enters a ______________, the colors are bent different amounts.
14. We see objects when light ______________ off them.
15. An apple reflects red light, and ______________ most of the other colors.
16. Tiny raindrops in the air can act like a prism to form a ______________.

Name______________________________ Date______________

Interpret Illustrations

Lesson 5

How Does Light Travel?

Illustrations are pictures or drawings of objects, people, or places. You can study the details in an illustration to find out information. Captions tell about the illustration.

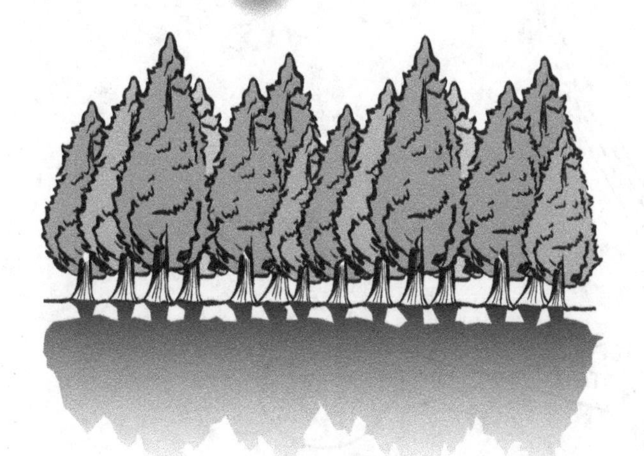

Rays of light travel in a straight line from the Sun to Earth.

Are these objects opaque?

Use the illustrations to answers the questions below.

1. What does the first illustration show?

____

2. What does the caption in the first illustration tell you?

____

3. What objects do you see in the second illustration?

____

4. Can the sunlight pass through the objects in the second illustration? How do you know? ______________________

5. How do you know that the objects in the second illustration are opaque?

____

Name ________________________ Date ____________

Interpret Illustrations
Lesson 5

Why Do You See Colors?

A diagram uses pictures or words to describe a thing or a process. The diagram shows how light is reflected off an object, allowing us to see colors.

Seeing an Apple

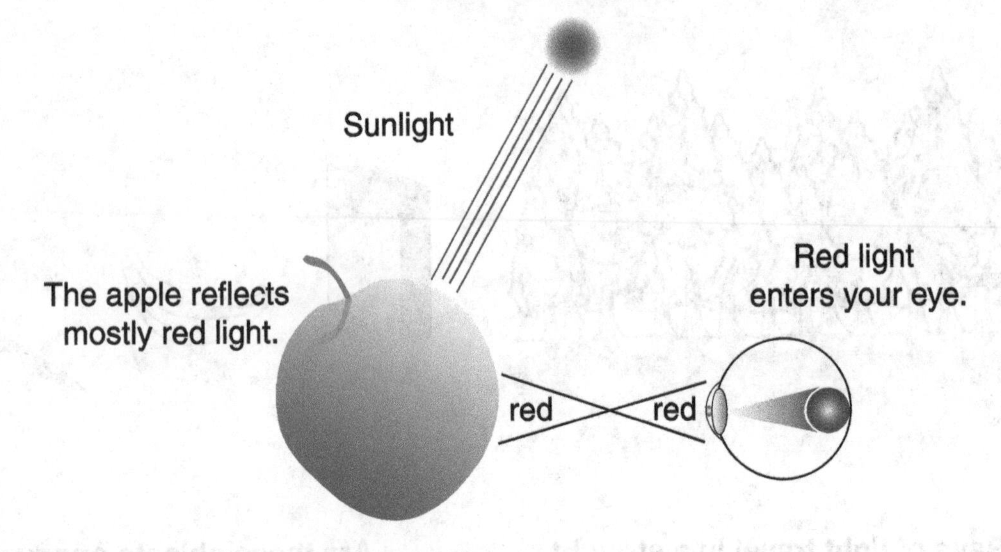

Use the diagram to answer the questions.

1. What is the title of the diagram?

____

2. Describe the path of light in the diagram. Where does it begin? Where does it end?

____

____

____

3. What color light is reflected off the apple? ____________________

4. What color light enters your eye? ____________________

5. What happens to the other colors that come from the Sun?

____

____

Name ______________________ Date ____________

Lesson Vocabulary
Lesson 5

How Light Travels

Fill in the blanks using words from the box.

Vocabulary
wood
tin can
water
plastic bottle
opaque
glass
reflects
cardboard
refracts

1. Light ______________, or bends, as it passes from air into water.
2. An ______________ material makes a shadow when light shines on it.
3. A curved mirror ______________ light in a different way from a flat mirror.

Complete each list using words from the box.

OPAQUE	NOT OPAQUE
4.	7.
5.	8.
6.	9.

Answer these questions in your own words.

10. You are on a camping trip, and you have lost your mirror. What other surfaces could you use to see your reflection? What do these surfaces have in common?

____

____

11. Is light a form of energy? What are natural sources of light? What kind of light is made by people?

____

____

____

Name ______________________ Date ____________

Cloze Test
Lesson 5

How Light Travels

Vocabulary

opaque	straight	shadows
natural	energy	

Fill in the blanks.

Light is a form of ________________. There are many different sources of light. The Sun, stars, lightning, and fire are examples of ________________ light sources. Light travels in ____________ lines from its source. Light passes through some materials but not through others. Materials that don't allow light to pass through them are ________________. Materials that block light create ________________.

Name_______________________ Date___________

Lesson Outline

Lesson 6

Properties of Sound

Fill in the blanks. Reading Skill: **Sequence of Events** - questions 4, 17, 18

How Are Sounds Made?

1. Sounds are made when objects ________________.
2. When a guitar string vibrates, it moves ________________ and ________________ quickly.
3. When an object vibrates, it makes the ________________ around it vibrate, too.
4. Sound moves to your ear after it moves through the ________________.
5. Sounds can travel through liquids such as ________________.
6. If you tap a ruler on the end of a desk, and put your ear to the other end of the desk, you can hear the ________________ noise through the desk.
7. Sound moves through metal better than it moves through nonmetal materials such as ________________.

How Do Sounds Get Higher or Lower?

8. A sound's ________________ is how high or low it is.
9. Sometimes, a change in pitch is caused by the ________________ or thickness of a string.
10. Shorter strings vibrate ________________ than longer strings.
11. The thinner the string, the ________________ the pitch of the sound.

Name ______________________ Date __________

Lesson Outline

Lesson 6

Fill in the blanks.

What Makes Sounds Loud or Soft?

12. The loudness or softness of a sound is called its ______________.
13. A loud sound takes more ______________ than a soft sound.
14. Whispering softly is a ______________ volume than calling out loudly.
15. The sounds of a space shuttle launch are ______________ than the sounds kittens make.

How Do You Hear Sounds?

16. A sound takes less than a ______________ to travel from across a room to your ear.
17. After a sound passes through your ear, a signal from the ear travels to your ______________.
18. Loud sounds can cause a loss of ______________.

Name ______________________ Date ____________

Interpret Illustrations

Lesson 6

How Are Sounds Made?

A diagram uses pictures and words to describe how something happens. Captions give you information about the pictures. This diagram has information that tells about how sounds are made.

When someone plays a trumpet, air vibrates inside it. Touch the trumpet and you can feel it vibrate.

Use the diagram to answer the following questions.

1. What does the picture show?

____

2. What do the waves in the picture represent?

____

3. What happens to the air inside a trumpet when someone plays it?

____

4. What happens when the air inside the trumpet vibrates?

____

5. How can you tell the trumpet is making a sound?

____

Name ______________________ Date __________

Interpret Illustrations

Lesson 6

How Do You Hear Sounds?

A diagram uses pictures and labels to show the sequence of events. The diagram below shows how sound travels through the ear to the brain.

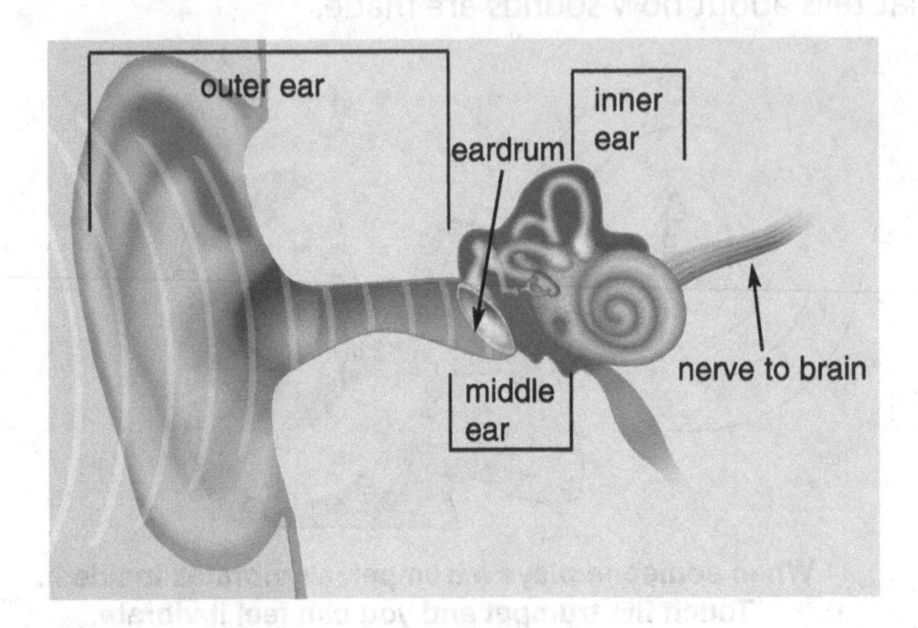

Use the diagram to answer the questions.

1. What three parts of the ear does the picture show?

____

2. What part of the ear does the sound first hit?

____

3. How does the shape of the ear help us hear?

____

4. What separates the outer ear from the middle ear?

____

5. What connects the bones of the inner ear to the brain?

____

Name______________________ Date__________

Lesson Vocabulary

Lesson 6

The Properties of Sound

Fill in the blanks.

Vocabulary
vibrate
pitch
volume
particles
energy
signal
outer ear

1. In order to make a sound, an object must ________________.
2. ________________ in the air vibrate, allowing sound to travel.
3. When a sound becomes lower or higher, there is a change in ________________.
4. Sounds that are very loud use more ________________ than sounds that are very soft.
5. The sound of a jet plane has a much higher ________________ than the sound of a car.
6. Sound first hits the ________________, then travels to the eardrum.
7. After sound travels through the ear, it sends a ________________ to the brain.

Answer each question.

8. How does the length of a string affect the pitch of the sound it makes?

____

____

9. What are some objects through which sound can travel?

____

Name_____________________________ Date_______________

Cloze Test
Lesson 6

The Properties of Sound

Vocabulary

air	liquids	solids
particles	faster	vibrate

Fill in the blanks.

Sound can travel through _______________ such as water. Sound can also travel through _______________ and gases. When an object vibrates, _______________ in the _______________ move back and forth around the object. Sounds get higher or lower in pitch when they _______________ faster or slower. The ___________ a string vibrates, the higher the pitch of the sound.

Name________________________________ Date______________

Lesson Outline

Lesson 7

Paths for Electricity

Fill in the blanks.

What Makes a Bulb Light?

1. A cell is a source of ________________.
2. A ________________ is made up of two or more cells.
3. You can use a bulb, wire, and cell to ________________ the bulb.
4. Electricity is a form of ________________ that travels in a circuit.
5. The path that electricity flows through is called a(n) ________________.
6. For a bulb to light, wires must ________________ it to the cell to form a circuit.
7. Electricity flowing through a circuit is called ________________.
8. Electricity must flow through a complete path, which is called a(n) ________________.
9. A ________________ opens or closes an electric circuit.
10. Turning a switch on allows electricity to ________________ in a complete path.
11. Electricity cannot flow through a(n) ________________ circuit.

Name ______________________ Date ____________

Fill in the blanks.

How Do You Use Electricity?

12. Many homes use ________________ lights.
13. Many people use ________________ to heat their homes.
14. Refrigerators, vacuum cleaners, and hair dryers are all electric ________________.
15. Radios and computers both run on ________________.
16. As electricity flows through a coil inside a toaster, it changes to ________________.
17. ________________ lights the tiny dots that make up the picture on a computer screen.
18. Electricity runs the ________________ inside a vacuum cleaner, which causes the vacuum to clean the carpet.

Name______________________________ Date______________

Interpret Illustrations

Lesson 7

What Makes a Bulb Light?

A diagram is a way to organize information. This diagram compares open and closed circuits.

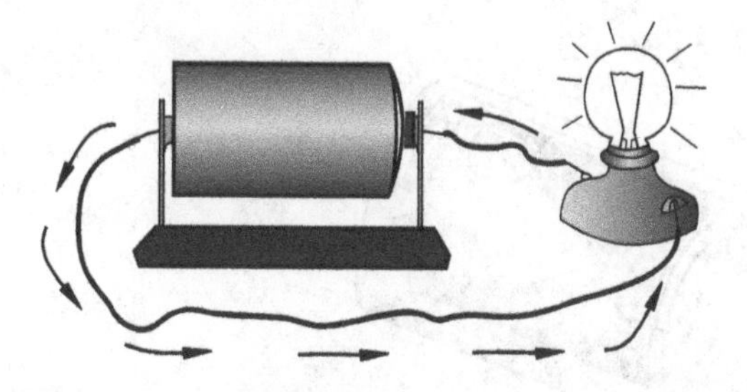

Electricity flows.

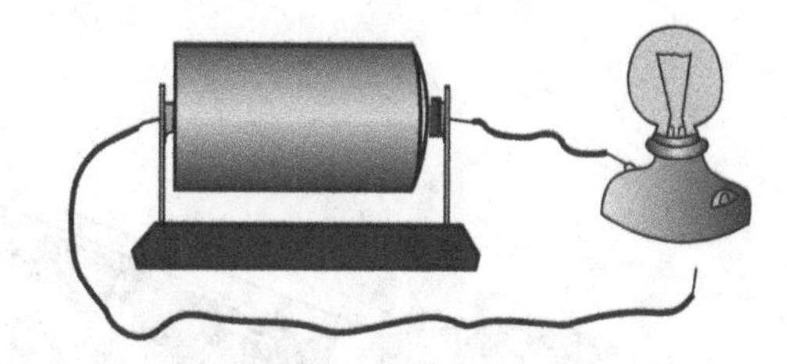

Electricity does not flow.

Use the diagram to answer the following questions.

1. Which picture shows an open circuit? How can you tell?

____

____

2. Can electricity flow through the open circuit? Explain.

____

____

3. Which picture shows a closed circuit? How can you tell?

____

____

4. What would stop the flow of electricity through the closed circuit?

____

____

5. What might you do to get the electricity to flow in the open circuit?

____

____

Name______________________ Date______________

Interpret Illustrations

Lesson 7

How Does a Flashlight Work?

Diagrams often show a process or how something works. The diagram below shows the path of electricity through the inside of a flashlight.

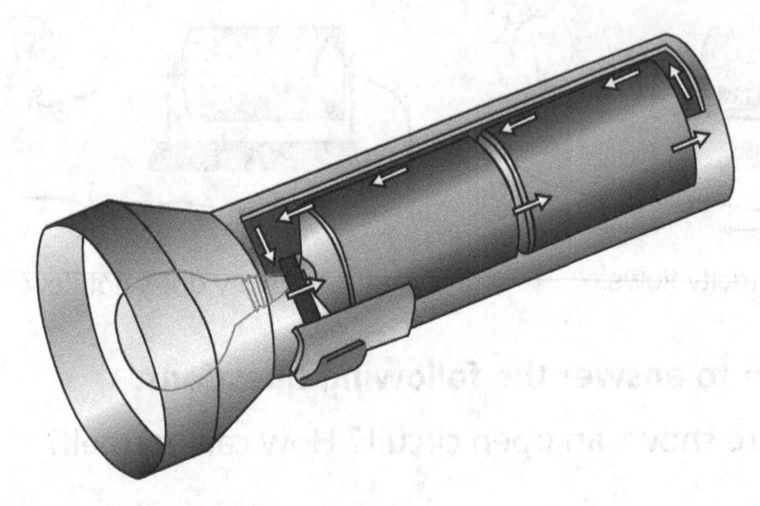

Use the diagram to answer each question.

1. How would you be able to turn on the flashlight?

____

2. What is shown by the arrows in the diagram?

____

3. Where does the electricity come from?

____

4. If the switch is off, what happens to the circuit?

____

5. When the switch is turned on, what happens to the path of electricity?

____

Use with textbook page F73

Name________________________________ Date______________

Lesson Vocabulary
Lesson 7

Paths for Electricity

Use the words in the box to complete sentences 1–6.

Vocabulary
open
switch
electric current
circuit
closed
cell

1. A path that electricity flows through is called a(n) ______________.
2. Electricity flowing through a circuit is called a(n) ______________.
3. Electricity will not flow through a circuit that is ______________.
4. A device that is used to open or close an electrical circuit is called a(n) ______________.
5. You can play your CD player when its electrical circuit is ______________.
6. One source of electricity is called a(n) ______________.

Answer these questions in your own words.

7. Describe the closed circuit in a flashlight. Is the flashlight on or off when the circuit is closed?

____

____

____

8. A storm has caused the electricity in your home to go off. Which items that use electricity might still work? Why?

____

____

____

Name________________________ Date____________

Cloze Test
Lesson 7

Paths for Electricity

Vocabulary

cell	path	light
energy	current	circuit

Fill in the blanks.

Electricity is a form of________________ that travels in a(n) __________. A circuit is a(n) ________________ in which electricity goes around and around. For the bulb to ________________, wires must connect the bulb to the ________________. Electricity flowing through a circuit is called electrical ________________.

Name______________________________ Date______________

Chapter Vocabulary
Chapter 12

Energy

Circle the letter of the best answer.

1. A sound is made when an object
   a. heats. b. refracts.
   c. vibrates. d. conducts.

2. Heat travels easily through a conductor, but does NOT travel easily through a(n)
   a. insulator. b. thermometer.
   c. solid. d. liquid.

3. Heat always flows from warmer objects to
   a. the ground. b. hot objects.
   c. plants. d. cooler objects.

4. Opaque items, which light cannot pass through, create
   a. darkness. b. shadows.
   c. energy. d. matter.

5. A prism is a thick piece of glass that
   a. creates a shadow. b. refracts light.
   c. creates energy. d. is reflected off the object.

6. A cell is
   a. a source of electricity. b. a kind of circuit.
   c. a form of energy. d. something that opens and closes a circuit.

Name____________________________ Date____________

Chapter Vocabulary

Chapter 12

Circle the letter of the best answer.

7. When light rays pass from one material to another, they

a. disappear. **b.** break.

c. bend. **d.** make shadows.

8. The path that electricity follows is called a

a. cell. **b.** circuit.

c. switch. **d.** light.

9. When matter is heated, it

a. expands. **b.** changes form.

c. changes color. **d.** gets smaller.

10. Which of these items is NOT opaque?

a. a computer disk **b.** a window

c. an apple **d.** a movie screen

11. Heat is

a. a measure of temperature.

b. a mixture.

c. a form of electricity.

d. a form of energy.

12. A closed circuit

a. prevents electricity from flowing.

b. means you have a dead light bulb.

c. allows electricity to flow.

d. makes the temperature rise.

Name_________________________ Date______________

Words and Meanings

Draw a line from each word to its meaning.

Words	**Meanings**
1. switch	a. matter that has shape and volume
2. element	b. characteristic or trait
3. mass	c. matter that has volume but not shape
4. compound	d. matter with no definite shape or volume
5. solid	e. how much space matter takes up
6. liquid	f. building block of matter
7. property	g. how much matter is in an object
8. gas	h. two or more elements put together
9. volume	i. opens or closes an electric circuit

Now write the missing word or meaning below.

Words	**Meanings**
______________	how hot or cold something is
heat	______________
______________	doesn't let light pass through it
reflect	______________
______________	path electricity flows through

Name ______________________ Date ____________

Unit Vocabulary

Unit F

Crossword

Read each clue. Write the answer.

Across

3. Attracts iron
5. Has volume but no shape
8. Building block of matter
11. Doesn't let light through
12. Has no shape or volume
13. Has definite shape and volume

Down

1. Unit for measuring temperature
2. Shiny material from the ground
3. Different matter mixed together
4. Contains one kind of matter spread evenly throughout
6. How much space something takes up
7. Two or more elements together
9. Warms things up
10. How much matter is in an object

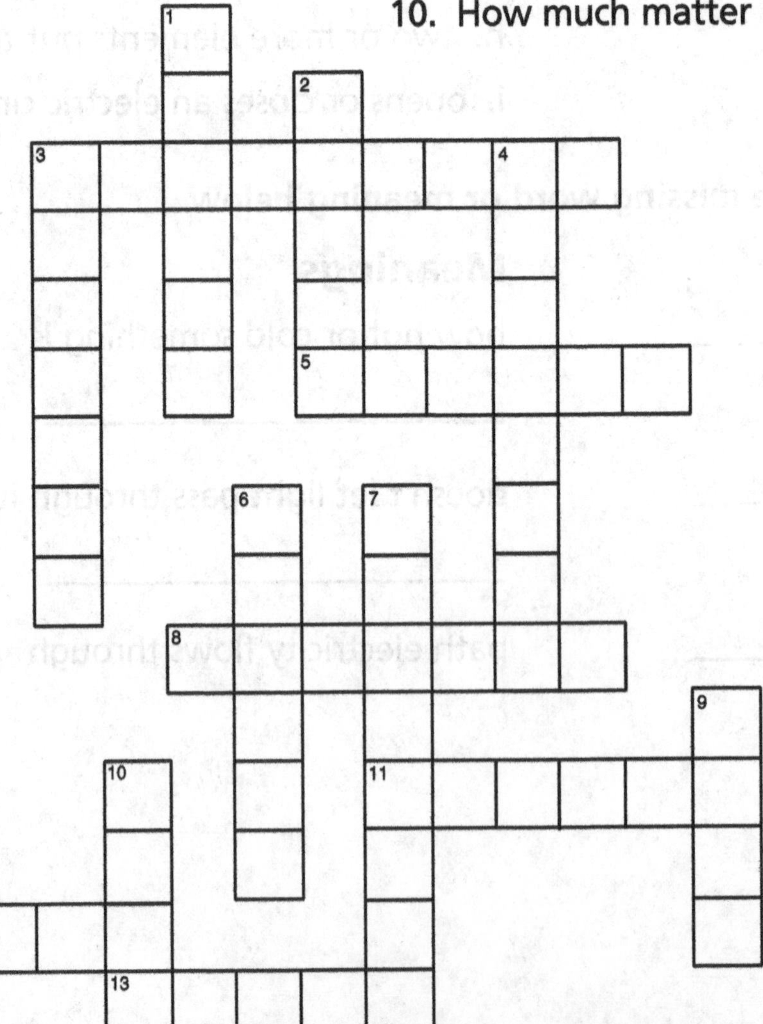

Name________________________________ Date______________

Unit Vocabulary
Unit F

Word Wheel

Start with the letter E. Write that letter on the first line below the puzzle. Now write down every third letter on the rest of the lines. (The next letter has a star next to it to help you.) Continue around the wheel, writing down every third letter. After you go around three times, you'll have an important message.

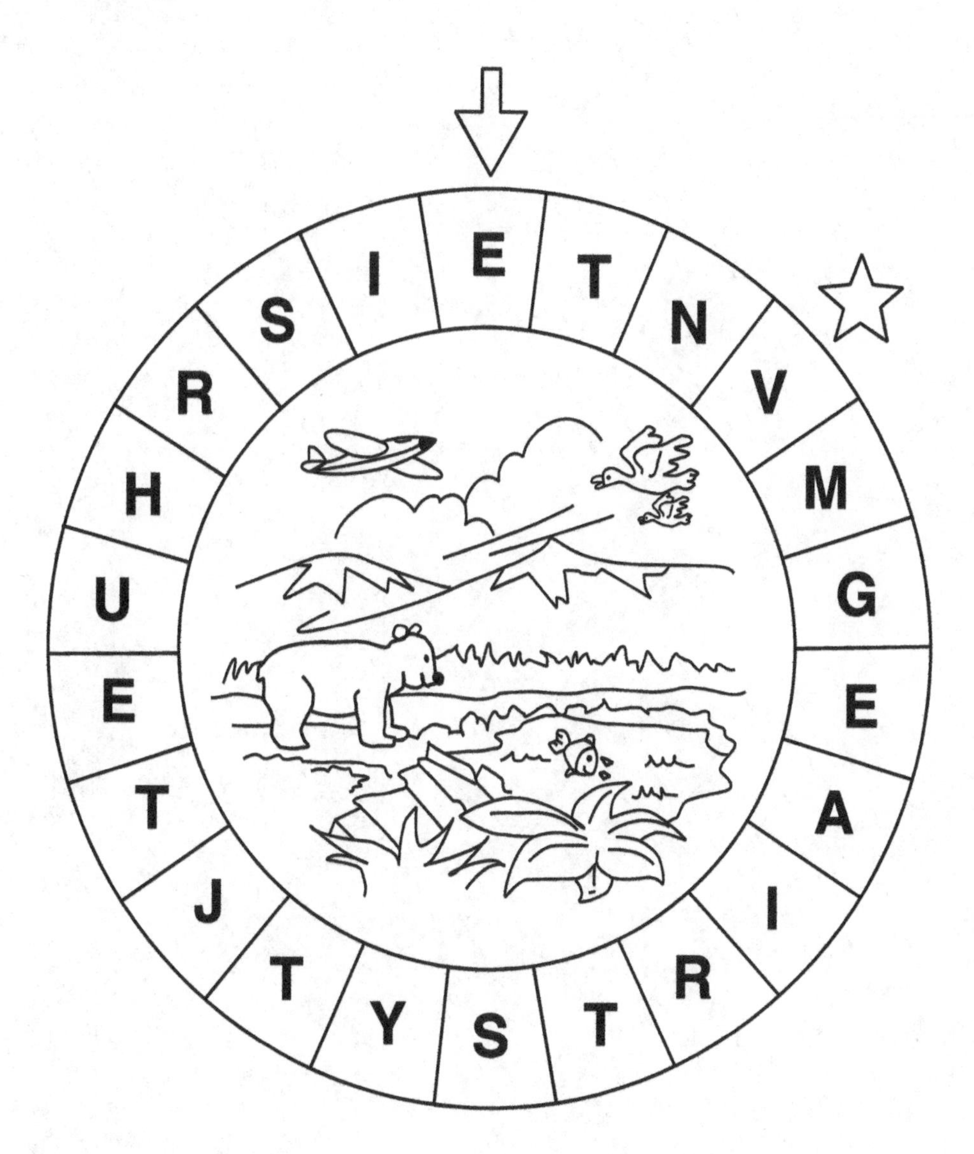

___ ___ ___ ___ ___ ___ ___ ___ ___ ___ ___ ___ ___   ___ ___

___ ___ ___ ___ ___   ___ ___ ___ ___ ___ ___ ___!